GENDER

WHAT EVERYONE NEEDS TO KNOW®

GENDER

WHAT EVERYONE NEEDS TO KNOW®

LAURA ERICKSON-SCHROTH AND
BENJAMIN DAVIS

OXFORD
UNIVERSITY PRESS

OXFORD
UNIVERSITY PRESS

Oxford University Press is a department of the University of Oxford. It furthers the University's objective of excellence in research, scholarship, and education by publishing worldwide. Oxford is a registered trade mark of Oxford University Press in the UK and certain other countries.

"What Everyone Needs to Know" is a registered trademark of Oxford University Press.

Published in the United States of America by Oxford University Press
198 Madison Avenue, New York, NY 10016, United States of America.

Library of Congress Cataloging-in-Publication Data
Names: Erickson-Schroth, Laura, author. | Davis, Benjamin, author.
Title: Gender : what everyone needs to know /
Laura Erickson-Schroth & Benjamin Davis.
Description: New York, NY : Oxford University Press, [2021] |
Series: What everyone needs to know |
Includes bibliographical references and index.
Identifiers: LCCN 2020020674 (print) | LCCN 2020020675 (ebook) |
ISBN 9780190880033 (hardback) | ISBN 9780190880026 (paperback) |
ISBN 9780190880057 (ebook)
Subjects: LCSH: Gender identity. | Gender expression. |
Gender nonconformity. | Sex role. | Sex differences.
Classification: LCC HQ18.55 .E75 2021 (print) |
LCC HQ18.55 (ebook) | DDC 305.3—dc23
LC record available at https://lccn.loc.gov/2020020674
LC ebook record available at https://lccn.loc.gov/2020020675

1 3 5 7 9 8 6 4 2

Paperback printed by LSC Communications, United States of America
Hardback printed by Bridgeport National Bindery, Inc., United States of America

CONTENTS

ACKNOWLEDGMENTS

First of all, I'd like to thank my co-author, Ben, who is not only a great writer but a wonderful human being. Making the decision to work on a project like this with a friend was a brave one for both of us. There were moments when I had to apologize profusely for missing deadlines, and somehow he forgave me. I love him more than I did before we started, which I didn't think was possible.

My mother, Nancy S. Erickson, JD, LLM, MA, was immensely helpful during this project. She is a lawyer with extensive knowledge of women's legal history and has been editing my writing since I learned how to write. In addition to providing feedback on parts of this book, she also wrote the first draft of the section on women's legal rights.

Finally, thank you to Abby, who is a true partner to me in every sense of the word. I don't think there is anything in life we can't accomplish together.

—Laura

I owe tremendous gratitude to Laura, my co-author. This book was her vision, and I am humbled to have been included in it. Laura's incredible knowledge and clarity is certainly what allowed this book to make it to print. Thank you, Laura, for your brilliance, humor, and dedication.

Thank you to Clare, Maddie, Caty, Ember, Julia, Jess, Beck, Annie, and Todd, who provided endless encouragement and love, always.

To transgender and gender diverse people across the globe who demonstrate daily the beauty in difference, thank you.

—Ben

INTRODUCTION

Gender is all around us. Beliefs about gender impact our jobs, families, schools, religions, laws, politics, relationships, sports, clothes, and so much more. Gender permeates almost every aspect of our lives as humans.

Although this book is part of a series called "What Everyone Needs to Know," it would be impossible to cover everything known about gender in one book, and since gender is something we all have in common and at the same time all experience differently, a consensus on the "most important" parts of gender differs based on personal experience and interest. In this book, we've tried to give you the highlights, so that you can dig deeper on your own if you hit a topic that's interesting to you.

With a book like this, there is always the question of what to include. Given space limitations, by making a decision to cover a topic, we have also made a choice not to cover another topic. And who are we to decide? Inevitably, we have left things out that authors from a different social or political background might have included.

Because gender is omnipresent, it is intertwined with so many other facets of our identities and lives. We cannot talk about gender without talking about race or class, for instance. Although we are not the experts on these intersections, we have attempted to identify the most crucial conversations so that readers are aware that they are happening.

Gender, and the words we used to describe it, depend on where we live and who we interact with. They are also constantly evolving. Still, there are terms that are important to know to have a common language to start from.

Gender is an individual and social experience, as opposed to *sex,* which is determined by chromosomes and hormones. To complicate things, there are overlaps between these two concepts, and there is evidence that gender may be biologically influenced. An individual's sex may be male, female, or intersex. Those who are intersex may have chromosomes or hormones that vary from expected binary combinations.

Gender identity is a person's inner sense of their gender as male, female, or something else. *Gender roles* reflect societal expectations for behaviors based on gender. A person's *gender expression* involves their mode of demonstrating gender to the world, through clothing, hair style, and mannerisms.

Transgender, or *trans*, people are those whose gender identities are different from their genders assigned at birth. Those who are *cisgender* have gender identities that match their assigned genders. Some people identify as *nonbinary,* meaning that they do not see themselves as either a man or a woman, but something outside of or in-between these. Binary transgender people typically use traditional *pronouns* such as he/him or she/her, while nonbinary individuals may use other pronouns, including they/them or ze/hir.

Sexual orientation is separate from gender identity and reflects the genders or sexes of the people someone is attracted to. The acronym *LGBTQ* is typically expanded as lesbian, gay, bisexual, transgender, and queer or questioning and may have additional letters added depending on the situation.

Lesbians are those who identify as women and are attracted to other women. *Gay* is a more gender-neutral term and can apply to anyone who is attracted to those with the same gender identity as themselves. *Straight* people are men attracted to women, or vice versa. *Bisexual* refers to those who are attracted to more than one gender. *Pansexual* is becoming a more popular term to describe similar attractions and, for some, signifies a less binary approach to sexuality. There are also those who identify as *asexual* and may have relationships but do not feel sexually attracted to others.

Queer is a complicated term that was reclaimed after originally being used as a slur. In some contexts, it is political—signifying a resistance to traditional expectations—while in other situations, it is more of an umbrella term to describe those who are not straight, and sometimes, those who are not cisgender.

Gender roles across time and throughout the world are varied and diverse, although most societies are *patriarchal*, elevating men above women by giving them more power. Patriarchal systems are typically based in *sexism*, discrimination based on gender. Most cultures are also *heterosexist*, biased in favor of *heterosexual* (straight) people; *heteronormative*, centered around straight culture; and *homophobic*, biased against *homosexual* (gay) people. They are also *cissexist*, biased in favor of cisgender people; *cisnormative*, centered around cisgender culture; and *transphobic*, biased against transgender people.

The United States, along with many other countries, has a history of colonization, where one group has come in by force and taken over another, affecting every aspect of the colonized community's existence, including their gendered lives. Many in the United States face oppression based on more than one aspect of their identities. Oppression can also weave together with *privilege*, which is the advantage afforded to certain groups based on identities such as gender, race, class, religion, and ability. *Intersectionality* is an approach that allows us to explore these experiences, keeping in mind the many interdependent systems that affect our lives.

This is not an exhaustive list of terms, and those included here are bound to change over time, but they may allow you to familiarize yourself with key concepts related to gender as you move forward to the pages that follow. It is our hope that this book inspires you to explore the world of gender wherever it may take you.

1

GENDER 101

How is gender different from sex?

Gender and *sex* are terms that are often used interchangeably and can have different meanings in different contexts. However, in general, sex refers to physical characteristics, and gender to social aspects of identity.

Sex is usually assigned at birth as either male or female depending on the appearance of the genitals. There are also many people who are intersex, meaning that their bodies do not fully match our expectations of either male or female. Intersex people may be identified at birth, but many are not.

We have certain ideas about what "typical" sex development looks like. Those with XX chromosomes are expected to be born with ovaries, a vagina, a vulva, and a clitoris. When they reach puberty, we assume they will develop breasts and start to menstruate. Those with XY chromosomes are expected to be born with testicles and a penis. When they reach puberty, we assume they will grow tall, develop facial and body hair, and produce sperm.

Even among those who match these typical trajectories, there can be a tremendous amount of variation. There are many tall women and short men. There are women with lots of body hair and men with very little. Intersex people can differ from these expectations in many ways. They may have XY chromosomes and a vagina, or XX chromosomes and an enlarged clitoris. Sex may seem simple compared to gender, but it's more complicated than we think.

Just as we are assigned a sex at birth, we are also assigned a gender, and our assigned gender generally reflects our assigned sex. Those born with a vulva and a vagina are typically raised as girls, and those born with a penis and testicles are typically raised as boys. Gender roles—the positions and responsibilities expected of us based on our assigned gender—vary from country to country and even within countries depending on our social groups. Clothing considered "normal" for women or men in one country may seem extremely unusual in another. The same is true of behaviors and relationships.

Some people talk about gender as a "social construction," something that is created by society. Given such wide variation in gender expression—appearance, clothing, and behavior—across the world, it is clear that many aspects of gender are learned. However, there is controversy over the relative contributions of "nature" and "nurture" to our gender identity—our internal sense of our own gender.

While gender roles and gender expression may depend heavily on social influences, there is building evidence that gender identity may have biological components, such as genetic or hormonal influences. Prenatal hormone exposure appears to have an influence on gender identity, although in ways we have yet to completely understand. If gender identity is in some part biologically based, then the line between sex and gender becomes even more complex.

What is gender identity?

Gender identity is our self-conception of who we are—our innermost sense of being a man, a woman, or something else entirely. For some people, gender identity is consistent with their sex assigned at birth. People who have penises often unquestionably identify as men, and those with vulvas, as women.

But what makes someone a man or a woman? What is gender identity, really? If a woman undergoes a hysterectomy (removal of the uterus), most would agree that this does not mean she is no longer a woman. If a man needs one or both testicles removed, he may grapple with what this means for his masculinity, but he remains a man.

Even without certain body parts, there is something that connects us to our womanhood or manhood, something else that identifies us with gender—even when we can't see, don't know, or have lost an aspect of the body that typically confirms sex. That "something else" can be thought of as gender identity—how our brain thinks about ourselves as gendered beings, and what those genders mean to us.

Gender identity is complex and varies with cultural context. Our gender identities are shaped by the fact that we grow up in different homes and families, with distinct traditions and gendered role models. *Man* and *woman* are the words we use to identify a complex array of gendered thoughts, ideals, and assumptions.

Most infants with penises grow into self-identified boys and men, who have similar ideas about what that means. The name we use for this group of people is cisgender, meaning that there is a congruence between their sex assigned at birth and the gender with which they identify. Transgender people, on the other hand, identify with a gender different from the one they were assigned. Cisnormativity, the cultural assumption that people are going to be cisgender, influences the way we see ourselves growing up.

Cisgender people, when asked how they knew they were a boy or a girl, often respond as if the question were preposterous. Many transgender people similarly discuss their gender as being forever constant and intrinsically known—that despite the genitals between their legs, the distinct feeling of being a man, woman, boy, or girl, was always there. By three years old, most toddlers have a sense of their gender, and soon after, some children whose gender identities do not match societal expectations are able to identify this discrepancy.

Gender identity forms early on as part of a developmental process of identity formation wherein a young person begins to understand themselves and their place in their world. It is important to note that not all children have the language, sense of safety, and support they need to talk freely about their gender. Someone feeling like their gender and sex are incongruent may not share that information through their words or actions as a child or even as an adult.

Because many children seem to have a sense of their gender identity very early on, it is likely that gender identity is biologically influenced, although biology does not appear to be the sole determinant of gender identity. There is growing evidence connecting gender identity to prenatal hormone exposure, although there is much debate about where, why, and how gender identity is formed.

Language around gender identity is specific to generation, location, and cultural context. And language is constantly evolving. The words one person uses to identify may be offensive to someone else. It is always best to follow a person's lead when talking about their gender identity, working to understand the words and language they use to describe their gender most accurately, and what those words mean to them.

What is gender expression?

Gender expression is a term used to describe a person's outward appearance. Characteristics that are commonly gendered include clothing, jewelry, and hair length and style. Gender expression also includes activities, interests, and mannerisms that are observable. Acting tough or sporty or being aggressive, dominant, or unemotional are all qualities associated with masculinity in a western context. Behaving in a nurturing way, or being gentle, are seen as feminine. These traits are observable through action and may or may not correspond with an internal identification. There are people whose gender identities are more feminine, but whose gender expressions are more masculine, and vice versa. Gender expression may also change over a person's lifespan. Young people often present in hyperfeminine or hypermasculine ways, and their gender expression often becomes less stereotypical as they come to understand themselves and their own identities better.

Masculinity is often seen as corresponding with confidence, which is signaled through eye contact, bold body language, and a loud voice commandeering attention. We have come to know these attributes as those that portray credence and conviction. Similarly, we understand masculinity to be presented to us via a series of postures. Stoicism, assertiveness, wearing neutral or dark colors, having shorter hair styles, and possessing courage and determination all indicate masculinity.

Femininity is signaled through a person's tendency to emote, accessorize, be more delicate in their mannerisms and dress, show more of their body through their clothing, wear makeup, and engage in less competitive, more nurturing activities. Wearing skirts and dresses and donning the color pink are all expected feminine ways to express gender. Buying blue clothing for a baby boy and, later, trucks and sports equipment, evidence an overwhelming assumption that boys will embody and choose to portray masculinity in a very specific way.

Outward expression can be significantly different depending on the cultural context. For example, in the United States, eye contact is expected and positively regarded and is taught in schools to increase effective conversation, persuasiveness, and as an indicator of truth and understanding. However, in China and Japan, eye contact can be seen in certain contexts as insubordinate or disrespectful. In Latin American and African communities, eye contact can, at times, be perceived as aggressive, and in the Middle East, the same quality of eye contact could be seen as a romantic gesture.

The ways in which masculinity and femininity are demonstrated and perceived across culture and time vary significantly as well. Carrying ornate decorative accessories, adorning jewelry, and wearing makeup have all aligned with power and masculinity in some areas of the world and times in history but would be wildly out of place if the goal was to demonstrate masculinity in the United States today.

It can be surprising to observe the rate at which gendered characteristics of expression are continuously reorganized. Crying, which in our culture indicates vulnerability and is linked with femininity, was a sign of heroism and strength in the feudal period of Japan, throughout the Middle Ages, and in ancient Greece. Makeup, high heels, stockings, and wigs signaled power and masculininity in western countries in the 17th and 18th centuries. Even the color pink, now avoided by those invested in maintaining a masculine gender expression, was considered a strong and passionate color, more suitable for a boy, until World War II.

Gender expression is a complex series of actions that serve to prompt those around us to better understand who we are inside. However, the strength of cultural norms dictating what expressive qualities and characteristics are appropriate to indicate masculinity and femininity have reinforced a narrow and binary range of acceptable behaviors for men and women, respectively. Recently, attention has been drawn to the potentially harmful ramifications of such a strictly gendered stratification. In the 2017 Global Early Adolescent Study, rigid gender expectations were shown to increase the risk of girls entering child marriages, becoming pregnant as teens, and experiencing intimate partner violence. Boys in areas with more rigid gender expectations had elevated rates of substance abuse and suicidality.

Some progress is being made in expanding the ways in which both children and adults are permitted to express themselves. In 2015, the retail store Target announced that it was integrating toys from the boys' and girls' aisles, and a number of companies have begun to produce gender neutral clothing. As gender roles change and expand, so does the way culture creates observable markers, or expressive elements, of personal identity.

What are gender roles?

Each community expects women and men to think, feel, and act in certain ways, simply because they are women or men. Community norms, whether implicit or explicit, shape gender roles, and these expectations are often deeply woven into the fabric of daily life. Coined in 1955 by sexologist and psychologist John Money, the term *gender roles* describes those spoken and unspoken duties that are assigned to a person based on their sex or gender.

Gender roles not only vary throughout the globe and over time but also within a particular community. Gender roles differ for youth and adults, with adolescence often serving as a pivotal point in a young person's awareness and ownership over the traits and characteristics they embody. Gender roles may differ by profession and within families, may change from one side of the country to another, and may shift by the day as cultural icons popularize gendered characteristics.

Why are gender roles so important to us? What about the human experience necessitates a desire to categorize and reject that which does not fit?

Gender roles are fluid. Girls and women are often taught to be demure, modest, delicate, and subservient. Boys and men are often taught to be strong and confident, able to provide and protect. Over time we have expanded the narrow expectations for gender roles, although most of us continue to perform versions of masculinity and femininity, with care not to stray too far from the traditional.

Gender roles in the United States have changed significantly over time. The 1870 Census, which was the first in which women were counted in the workforce, shows that women were not only employed in factories and in education, but also worked as miners and steelworkers, ship riggers, gun and locksmiths, hunters and trappers, lawyers, dentists, and doctors. Historically, cultures have varied immensely in terms of permissible gender roles. Ancient Egyptian women could marry and divorce freely, work, and own property. Hatshepsut, Nefertiti, and Cleopatra were all powerful rulers in ancient Egypt, yet women fail to hold the highest positions of leadership today in the United States.

A cursory glance at gender throughout history and across the globe depicts gender roles as being wide and varied and largely assumed as "normal" by way of cultural significance rather than innate predisposition. Yet, despite acknowledged variance, gender roles tend to be strictly defined and reinforced. State regulated or not, a departure from the established regulatory order of gender can trigger outrage. The penalty can range from social ostracization to death.

As one strays from culturally accepted gender roles, a common experience of skepticism or devaluing occurs. Men who appear too feminine become comical in the public eye, and women who possess qualities aligned with masculinity can be ridiculed for being bitchy or bossy, overemotional, hysterical, and frigid. When expression moves away from an expected gender role, another common response is to question a person's sexuality. When a little boy is teased on the playground, for instance, and called a *faggot*, likely his tormentors have identified him as "acting like a girl," rather than having a conceptual understanding of that boy growing up to love and partner with other men. Physically strong women, women who find comfort in casual clothing or short hair, and women perceived as having aggressive qualities are similarly assumed to be queer, gay, or lesbian identified. In reality, we know nothing about the sexuality of the boy who likes pink and ballet, the athletic girl, or the female construction worker.

Over time, particularly throughout the course of the 20th century, gender roles have changed in a cultural shift toward being more gender equitable. Women are no longer solely expected to stay home with children, cook, clean, and accept a small subset of roles in the workplace. Numerous organizations dedicated to increasing the presence of women and girls in technology, mechanics, and medicine have yielded a more diverse workforce. However, women and men have hardly achieved economic parity. The 2017 Global Gender Gap Report, published by the World Economic Forum, estimated that the global economic gender gap would not be closed for another 217 years. In overall gender equality, encompassing four domains—economic participation and opportunity, educational attainment, health and survival, and political empowerment—the United States ranked 49th of 144 countries, similar to Serbia (40th), Uganda (45th), Botswana (46th), and Bangladesh (47th).

There are some signs of improvement in American women's gender equity. According to the Pew Research Center, in 46% of two-parent families, both parents work full time, up from 31% in 1970. In only 26% of these households does the father work full time and the mother is a homemaker. Still, statistics show that U.S. women spend about 4 hours a day on unpaid work, versus 2.5 hours for men. And, perhaps surprisingly, straight men aged 18 to 34 are no more likely than older men to participate equally in household labor. When it comes to parenting, women in the United States continue to do significantly more, with married women clocking in at an average of 13.7 hours of childcare per week, versus 7.2 hours for married men.

Some would argue that, while gender roles have expanded for women, allowing them to enter the workforce and engage more in civic life, men's roles have not similarly shifted. While men have begun to participate more in traditionally feminine arenas, there may not have been as dramatic a change in their ability to step outside of rigidly defined norms of masculinity. Because men's and women's roles are intimately linked, working toward more flexible gender roles for men may benefit people of all genders.

What is the gender binary?

The gender binary refers to a structure of understanding wherein sex and gender are viewed as consisting of only two choices—male/female, man/woman, masculine/feminine—with nothing in between. When we think about sex or gender in a binary system, we tend to identify characteristics that are exclusive to one or the other, rather than traits that many people across different genders share.

We can observe the gender binary everywhere. We are assigned "baby boy" or "baby girl" often before we are given our names and sometimes before we are born. In filling out a birth certificate, an application to drive, go to school, rent an apartment, or open a bank account, there are usually only two options to choose from: male or female. We are taught, both subtly and overtly, that men do things women cannot, women engage differently than men, and that there is a planetary divide between the sexes.

Inherent in the gender binary is power. In most modern cultures, men hold more power than women, and those who identify outside of the binary are often highly criticized, objectified, and targeted for violence. Gender fluidity in a world that is largely invested in a binary system of gender can be threatening and confusing because of existing power dynamics.

The gender binary is so ingrained in our culture that most of us do not notice or question it. Even the way we speak is inherently binary. In Spanish and other romance languages, words lack clarity without gendered pronouns and suffixes. In English, phrases like "he or she," "ladies and gentlemen," and "boys and girls" reinforce the seemingly unquestionable assumption that there are only two options.

Yet upon further inquiry, we find that throughout history there has been recognition of multiple genders. Native American communities identified individuals known as Two Spirits, and in 2014 India began officially recognizing Hijras, third-gender individuals who have been an active part of Indian culture for thousands of years. The Warias of modern day Indonesia and the Mahus of Hawaii are similarly nonbinary. Rabbi Elliot Kukla identifies six genders referenced in the Hebrew scriptures, all contributing varied and valued roles in daily life.

A significant number of people have begun to use words like *nonbinary* or *genderqueer* to describe themselves. They may see themselves as androgynous or gender neutral, or they may oppose the strict boundaries of binary gendering. As nonbinary identities become more accepted in our own culture, it is likely that all people, including those who identify with binary genders, will benefit from more expansive definitions of gender.

What is gender nonconformity?

Gender nonconformity refers to the experience of someone who does not identify or express the culturally accepted set of gendered behaviors or actions associated with the sex they were assigned at birth. There is a range of gender nonconforming behaviors. Gender nonconformity might describe a masculine woman or a feminine man, or someone whose gender identity is different from that which they were assigned at birth. Some even use the language gender nonconforming, or GNC for short, as the best-fit term to define their gender identity.

A gender nonconforming behavior means that the way someone is perceived transgresses the assumptions of how a man or woman should function in a certain setting. Someone may not be conforming to a typical gender role by choosing a profession that is typically associated with another gender. For instance, it could be considered gender nonconforming for a woman to pursue the fields of plumbing or garbage removal. Given the narrow reality of gender roles, straying outside the lines isn't actually that hard. One could say that both a woman buying her own car and a stay-at-home dad are not conforming to accepted gender norms.

Nearly everyone can recall a time when their behavior was not completely in line with gendered expectations. It may have been subtle. For a man, voicing emotions or crying could feel gender nonconforming. For a woman, negotiating a salary might feel far away from the way she was taught to behave in the workplace. After all, depending on the household, many children are not socialized, taught, or modeled these behaviors while growing up. Acting in a gender nonconforming manner can feel anxiety-provoking and make us uneasy. Moreover, behaving in a gender nonconforming way might surprise someone we interact with, making them anxious, embarrassed, or even angry.

The magnitude of *how much* someone presents as gender nonconforming often dictates the response they will receive, ranging from surprise and amusement to suspicion, hostility, and violence. A single father will likely be met with surprise but not hostility. The athletic girl may be mocked for being a "tomboy," but likely her role will not threaten her standing in her community. Both of these examples may even elevate the individual; they signal strengths across gender ideals and have come to be valued within certain portions of society.

On the other hand, the price one pays for behaving in a significantly gender nonconforming manner can be deadly. The gender ambiguous person often elicits a response of skepticism and concern. For many of us, not knowing someone's gender is difficult to sit with. We have a surprising level of investment in fitting people into boxes, in finding out what they *really are.* Gender nonconforming people historically have been seen as intentionally deceptive, those around them responding with acts of violence and aggression. Transgender people face constant scrutiny and far too often find themselves at risk of danger. If they visibly carry male and female characteristics, onlookers are tasked with "figuring out" how to categorize and process their presence. Those who come into contact with them may feel the need to interrogate their identity, weighing masculine and feminine traits to understand their bodies as falling into one of two distinct categories, and becoming angry and frustrated if they cannot. No one should suffer just for being who they are, and confusion should allow for curiosity and growth, never assault.

Unsurprisingly, gender nonconforming behaviors and characteristics are categorized as such based on culture and vary considerably based on the accepted gender roles in a particular time and place. One may get on a plane in one country wherein their presentation is in line with gender-based expectations and deboard in another country where they become an oddity. Dressed in a kilt, for instance, a Scottish man may leave his country with his masculinity intact. Upon arrival elsewhere, his kilt is interpreted as a skirt. His hairy legs seem shocking to his hemline, and he may become comical and bizarre.

Importantly, those who demonstrate gender nonconforming behavior, such as the kilted man or the stay-at-home dad, may identify in a variety of ways, including as cisgender, transgender, or nonbinary. Individuals who identify as GNC often see themselves as falling under a transgender umbrella. They may also use terms such as *nonbinary* (not identifying as either a man or a woman), *agender* (not having a gender), *bigender* (two genders), *multigender* (multiple genders), *genderfluid* (a fluid or changing gender identity), or *genderqueer* (another word to identify outside of the binary). Keep in mind that identity labels and umbrella terms are complex and changing. To respectfully use the correct gender identity for someone, it is usually best to ask them directly.

What does it mean to be transgender?

Typically, the term *transgender* refers to someone who identifies with a gender that is outside the expectations associated with the sex they were assigned at birth. Men who were born with vulvas and ovaries are commonly referred to as transgender men or trans men, and women who were born with penises and testes are usually called transgender women or trans women.

Being transgender is separate from identifying as gay, lesbian, or queer and should not be confused with sexuality or sexual orientation. Transgender is not an "extreme version" of being gay. In fact, many transgender people partner with those of a different gender and identify as heterosexual.

Transgender people come in all shapes and sizes, from all different backgrounds, races, religions, and cultures. Nearly all professional organizations agree that being transgender is not a result of trauma, poor parenting, or abuse. Some people believe that being transgender is a choice, but most experts in the field believe that the only choice is how and if to disclose the identity, not whether or not it exists. The decision to disclose, or "come out," as transgender varies from person to person.

Sometimes, a person is able to name their transgender identity very young. A toddler, for instance, might insist they are a boy or girl, baffling their parents who were told by a doctor, nurse, or midwife that the child was something else, usually based on the genital configuration they were born with. The child, born with a vulva, was identified as a baby girl. Yet the child's mind, despite body characteristics and genitals, is seemingly hardwired, knowing himself as male. For a young transgender girl, the opposite scenario may exist. Born with a penis, a young child is assumed to be a boy and treated as such until, contrary to expectations, she begins stating she is, in fact, a girl.

When children voice concerns about their assigned genders or display gender nonconforming behaviors, it often causes distress for caretakers. Parents might be concerned that their child is confused. Because transgender identity remains stigmatized in many communities, caretakers may assume they did something to cause the child to identify outside of gender norms.

Gender nonconforming children often experience cues from adults around them, signaling that harm or embarrassment could result from gender transgression. As adults, we know the anxiety of having to "come out," disclosing an aspect of one's identity not easily seen or read from observation alone. Young people may anticipate a negative reaction from their immediate community and make a decision to keep their gender identity secret. Sometimes, this process can unfold largely unconsciously, the mind compartmentalizing this element of self-understanding until the resources exist to sustain more full exploration.

Some transgender individuals describe their childhood very differently. They may say that they were less concerned with gender identity growing up; however, they had a strong sense of not fitting in or that something internal was either miscalibrated or mistuned. After all, we have limited language to describe gender incongruence. A young person may not have had the cognitive resources or frame to begin to discuss or even think about their gender identity with the language that could help clarify their experience to the adults around them.

At times, gender identity can be closely related to gender expression. A young person with a gender identity incongruent with the sex they were assigned at birth might be drawn to the gendered expectations and gender roles associated with the gender they do identify with. For instance, a young transgender boy may desire to look like other boys he knows and participate in activities and adopt mannerisms more aligned with masculinity. Before he discloses a transgender identity, he might be perceived as a tomboy.

However, many masculine-presenting girls, or tomboys, do not grow up to be transgender. Similarly, many feminine boys do not grow up to be trans girls. Likewise, not all transgender young people demonstrate gender-nonconforming behaviors. A transgender boy may be comfortable in skirts and dresses, despite feeling male inside. Transgender girls, often stereotyped as hyperfeminine, might feel comfortable presenting more masculine, despite their internal female identity.

So how do we know when a child is transgender? If feminine boys can grow up to be cisgender boys or transgender girls, and trans girls can express themselves in a masculine or feminine manner, how can we predict or understand the transgender experience? If a young trans boy likes dress-up and makeup but says he is a boy, can we tell him he is incorrect?

Transgender identities can be nuanced, and sometimes seen as contradictory to the observer. Traditionally, we associate a male sex assignment with an identity of a boy or man and a masculine expression. Similarly, a baby identified female at birth is normed to identify as a girl and then a woman, presenting feminine attributes and characteristics. When we consider transgender young people, the common narrative is that they are "born in the wrong body," with mismatched body and brain genders made obvious by consistent observable behaviors. However, this model links gender identity and gender expression, which are separate concepts. When we isolate sex from gender identity *and* gender expression, the experience of a trans person becomes varied. Moreover, the ability to tell who might be transgender and who is not becomes near impossible. So how do we know? We need to wait for the person to tell us. To date, this is the only reliable data we have.

Puberty for a transgender adolescent can be a time of great distress if they perceive their body moving in a direction away from their identified gender. Without intervention, a person will usually undergo a sex-specific puberty corresponding with the sex they were assigned at birth, sometime between the ages of 9 and 13. Those identified female at birth typically develop breasts, curves, body hair, and begin to menstruate. Those identified male at birth typically experience their voice deepening, testicles descending, body hair increasing, and the production of sperm. These changes can be devastating to some young people, and in consultation with a health-care provider, can be halted temporarily through the use of puberty-blocking medications, until the young person is able to make decisions about adult hormone treatment. Gender-affirming surgeries, if pursued later, often work to modify or undo some of the sex-specific pubertal changes, so as to help a person's experience in the world match more with how they identify.

Post puberty, a transgender person usually has to make an intentional decision about disclosing their transgender identity to their community. Depending on their presenting gender expression, the disclosure may confirm theories or suspicion or may come as an incredible shock. The experience of coming out often marks a stressful time in a transgender person's life. On one hand, this might be the time where an individual first shares their truest self. However, there is often concern for ostracization, blame, shame, or disownment.

Those who begin to identify as transgender later in life can have varied experiences. They may have known about their gender-diverse identities for years and suffered with this secret or may have lived full, rich lives, but with something always in the back of their minds. There is an infinite range of transgender experiences, particularly as it pertains to if, when, and how a person tells their story or changes their outward experience to match something deeper inside.

What does it mean to be intersex?

Intersex people are those whose sexual or reproductive organs develop differently than the typical male or female pathways. The antiquated term *hermaphrodite* is now considered offensive. Some physicians and researchers categorize intersex people as having "disorders of sex development." A less pathologizing phrase is "differences of sex development." Most intersex people are healthy, and their only major differences from others are in their sexual or reproductive organs. However, there are some intersex people who deal with medical conditions that come along with these differences.

Some intersex people consider themselves to be part of the LGBTQ spectrum, while others do not. Longer forms of the abbreviation often include the word *intersex* (i.e., LGBTQIA, which is sometimes expanded to mean lesbian, gay, bisexual, transgender, queer, intersex, and asexual).

Estimates as to the number of people who are intersex vary widely, from about two in every thousand to two in every hundred people. The smaller estimates are generally produced from chart reviews of the number of infants undergoing early genital surgeries, while larger estimates come from multiple sources and may include people who did not know they were intersex until they ran into reproductive issues in adulthood.

Intersex communities have long been stigmatized, and efforts have been made to identify and "fix" infants with ambiguous genitalia. Intersex people have been subject to poking and prodding by teams of doctors from a young age because of their status as "interesting cases," and some are subjected to early surgeries that can affect both fertility and sexual pleasure. Many intersex activists and supporters have spoken out against this type of pathologization of intersex bodies and have demanded the right to choose for themselves as adults whether they want to undergo any surgical changes. Their work has shifted the conversation somewhat, but there are still no laws in the United States banning genital surgeries on infants.

Parents, often along with advice from physicians and therapists, assign their intersex child a gender. The choice of gender can depend on many things, including the appearance of the genitals, but also on the cultural context. In the United States, for example, children with female-typical XX chromosomes born with a large clitoris are usually assigned female unless their genitals are extremely masculinized. In some Middle Eastern cultures, the same children are more likely to be assigned male. In U.S. culture, being a man with a small phallus is considered undesirable, while in some other cultures, it is more important to have a male child than exactly what that child's body looks like.

To understand more about common intersex identities, it can be helpful to review the process that guides the formation of our sexual and reproductive organs. Those who develop to have female appearing bodies typically have two X chromosomes (XX) and those with male appearing bodies typically have one X and one Y chromosome (XY). In utero, all embryos regardless of their chromosomal makeup have similar internal and external organ structures until about six to eight weeks into development.

Internally, we each start out with undifferentiated gonads that will become either testicles or ovaries. For those with a Y chromosome, a gene called SRY signals the gonads to develop into testicles, and then hormones, including testosterone, produced by the testicles, lead to the formation of male-typical internal organs, such as the epididymis and vas deferens. In the absence of an SRY gene, the gonads develop into ovaries and female-typical internal organs develop, including the uterus and fallopian tubes. A separate process governs our external sexual organs. Bodies without exposure to testosterone develop a vagina and labia, while those with testosterone and another hormone called dihydrotestosterone (DHT) develop a penis and scrotum.

There are many ways that bodies can be intersex. Starting from the chromosomes, there are people born with other configurations than either XX or XY. About one in every 500 people has Klinefelter syndrome, in which there are two X chromosomes and a Y chromosome (XXY). Those with Klinefelter usually appear male, with male external genitals, and often do not know that they have a chromosomal difference until they run into fertility issues. They may have less body hair and musculature, with more rounded hips and breast tissue. About half of people with Klinefelter produce sperm (although it may not be motile) and therefore have the possibility of fertility with the use of assisted reproductive technologies. Those with Klinefelter typically have male gender identities, although data suggest that there may be a slightly higher percentage than others assigned male at birth who transition and identify as transgender women.

There are intersex people who have differences in the formation of their gonads. Some may even have gonads that are part ovarian and part testicular tissue. Depending on the mix of hormones produced, their other internal and external organs may be more masculine or feminine. Gender identity in these people varies significantly depending on the amount of masculinization or feminization of their bodies in utero.

Hormones produced during development can have a significant effect on the external organs. Those who are XX but have congenital adrenal hyperplasia (CAH) have a blockage in one of the enzyme pathways that leads to hormone production, so their bodies are exposed to more testosterone than those of other XX fetuses. This can lead to masculinization of the external genitals, including anything from an enlarged clitoris to a large phallus and fused labia that appear to be a scrotum. Most XX people with CAH identify as female despite exposure to increased androgens.

Those with androgen insensitivity syndrome (AIS) develop in what some might call the "opposite" way from those with CAH. They may be born with XY chromosomes but have hormone receptors that are unable to respond to testosterone, so despite their XY chromosomes they can have female-appearing external genitals. There are different types of AIS, and some people have partial AIS, while others have complete AIS. Those with complete AIS typically, although not always, identify as female, while those with partial AIS have a wider range of gender identities.

In addition to testosterone, the hormone DHT helps to masculinize the external genitals. In XY people with 5-alpha reductase deficiency (5-ARD), the enzyme that converts testosterone to DHT is reduced, and similar to AIS, this prevents the development of male-typical external genitals. 5-ARD is prevalent in some areas of the Dominican Republic, where it is known as *guevedoce*, which translates to "testicles at twelve." This is because people with 5-ARD present with expected female external genitalia until puberty, when their testicles may descend and they can develop body hair, a deeper voice, and larger external genitals. Many people with 5-ARD are raised as girls but identify as male when they hit puberty. In cultures where 5-ARD is common, such as the Dominican Republic, Turkey, and Papua New Guinea, children with 5-ARD are sometimes identified early and raised as boys in anticipation of their physical changes at puberty.

Intersex people, like all of us, come in many shapes and sizes. Because of our tendency to see male and female as binary, many of us are unaware of how common it is to have bodies that do not match up 100% with what is expected. This can lead to early interventions such as genital surgeries on infants, and to lifetimes of stigma, when instead we could be celebrating the many different types of bodies that exist and encouraging each person to follow their own path.

Is gender identity based on nature or nurture?

The nature versus nurture debate about the origins of gender identity has been raging for decades and boils down to the relative contributions of biological and social factors. A significant amount of research has been done in this area, and most of it is in an attempt to demonstrate a biological basis for gender identity.

Biological theories are popular right now for a number of reasons. Although we like to think of science as objective, social structures affect the type of research that is done, by whom, and for what reasons. We live in an age where technology is moving at incredible speed, and there is public excitement over every new gene discovery. This translates into a research environment where there is much more funding for "hard" sciences than for social and psychological ones.

In the field of gender identity, there are also specific reasons that emphasis is being placed on biological research. Because of the stigma and discrimination that transgender people face, there is an effort to demonstrate that gender identity is biological and therefore not a choice that a person can make. In popular culture—including Lady Gaga's song "Born This Way"—similar arguments have been made to defend gay identities. In our current political environment, many transgender people and allies feel that this is the best way to advance transgender rights. If being transgender is not a choice, they argue, then we cannot blame someone for being transgender.

There are a number of flaws in this reasoning. Perhaps most frightening is the idea that a discovery of a particular biological "cause" of transgender identity could lead to the search for a "cure." This kind of breakthrough could also mean that anyone who does not possess a particular gene or other biological marker for transgender identity could have their identity questioned and could be considered to be pretending. Reliance on a biological basis for identity also sends a message to young people that being transgender, if it were a choice, would be a poor one.

Public understanding of gender identity as a biological phenomenon began in the 1990s with the publication of John Colapinto's book, *As Nature Made Him: The Boy Who Was Raised as a Girl*. Colapinto detailed the life of David Reimer, formerly known as John/Joan, who was born as a twin boy in Canada. As a baby, during his circumcision, Reimer had his penis nearly entirely destroyed. A well-known psychologist named John Money was brought in and recommended to the parents that they raise Reimer as a girl, arguing that a man without a penis would not be able to function as an adult and that children's gender identities were malleable if decisions were made at a young age.

Colapinto's exposé of the John/Joan case revealed that Reimer had not responded well to being raised as a girl. He eventually forced his parents to admit what had happened and transitioned to live as male. When this case was published, the public responded by wholeheartedly accepting arguments for the biological basis of gender identity, not understanding that gender identity is likely much more complicated than it was made out to be.

There are others like David Reimer who have been raised as girls despite having XY chromosomes and "typical" male hormone exposure as fetuses. Some have penile agenesis (the penis does not develop); others, micropenis (the penis is very small) or cloacal exstrophy (the intestines are exposed through the abdomen and often no genitals are clearly present). Approximately half of people in this situation eventually come to identify as male, despite being brought up as female. Still, half do not, suggesting a role for the social environment in the development of gender identity.

What is the current evidence that gender identity is influenced by biology? At the most basic level, we have our genes. Studies of chromosomes, the molecules that organize our genes, show that transgender people's chromosomes generally match their sex assigned at birth. That is, there is no known connection between chromosomal variations and transgender identity. There is also not yet any clear evidence that particular genes are involved in gender identity. A few small studies have suggested a possible link to hormone-related genes, but these studies have not been replicated.

Recently, researchers have attempted to better understand the basis of gender identity by using brain imaging technology. There have been a number of small studies comparing transgender and cisgender people who have undergone various forms of brain imaging. While there have been some "positive" results, meaning that differences were found, these have been difficult to replicate in other studies and their meaning is not yet clear. This is because finding distinctions in brain anatomy or function through magnetic resonance imaging, diffusion tensor imaging, or other types of brain imaging do not tell us *why* there are differences. Brain scans of people with posttraumatic stress disorder may be different from those of people without posttraumatic stress disorder, but these changes were influenced by a traumatic event. Not all biological differences are present before birth, and our bodies are constantly changing in response to our environments. If differences are found on brain scans between transgender and cisgender people, we do not know if they were there from birth or developed in response to our environments.

Researchers, starting in the 1970s, took blood samples from gay and transgender people, hoping to show that they had different levels of hormones than straight and cisgender people. However, no replicated studies have ever found differences in adult hormone levels between these groups. Scientists have, for the most part, given up this line of inquiry, although more recently there has been some debate about possible increased rates of polycystic ovary syndrome, characterized by increased testosterone levels, in trans men prior to starting prescribed testosterone. Studies have been small and contradictory.

Where there is some evidence of a possible biological influence on gender identity is in prenatal hormones—those hormones we are exposed to while still fetuses. However, we currently have no good way to measure this exposure, as it can be dangerous to the fetus to take the blood or tissue samples that would be needed. Animal researchers, on the other hand, have done experiments that would not be permitted with humans. These studies have shown that injecting testosterone into female rodent fetuses leads to a change in their behavior later as adults—specifically, an increase in their tendency to mount other rodents, which is considered more of a male-typical behavior. However, it can be hard to extrapolate from these findings to humans. Do the testosterone injections affect the rodents' gender identity or their sexuality? What is the meaning of mounting behavior? We can hardly ask the rodents themselves.

Intersex people—those whose bodies develop differently from typical male or female bodies—have also been the subjects of research into gender identity, many without their consent. Because certain differences of sex development come with changes in prenatal hormone exposure, we may be able to learn something about a possible connection between hormones and gender identity by studying how intersex people identify. There are many ways of being intersex, and each comes with a different hormonal exposure. When intersex children are born, they are generally assigned either a male or female gender based on the opinions of their doctors and parents.

People with certain differences of sex development are more likely than others to change their gender identification later in life. In 5-ARD, a person with XY chromosomes can be born female-appearing because of an inability for the body to convert testosterone to a more potent hormone that masculinizes the genitals. As people with 5-ARD go through puberty, however, they begin to appear more masculine, developing a deeper voice, growing male-pattern body and facial hair, and experiencing descension of testicles and enlarged genitals. People with 5-ARD often initially identify as girls and later as men. Some people argue that this is because they have been exposed to male-typical hormones as fetuses. However, there are a number of social and psychological reasons someone in this position might identify as male, not the least of which is how difficult it may be to live as female in a body that appears to be that of an adult male.

In congential adrenal hyperplasia (CAH), someone with XX chromosomes may be exposed to increased male-typical hormones as a fetus. People with XX chromosomes and CAH (sometimes called "CAH girls") can often be born with a larger than usual clitoris. There have been numerous studies of CAH girls, and some seem to show more male-typical ("rough and tumble") play as children, as well as higher rates of lesbian/bisexual or transgender male identity as adults. However, most CAH girls identify as cisgender straight women. It has yet to be shown exactly what influence prenatal hormones may have on the sexuality or gender identity of a growing fetus. Interestingly, current theories about the development of both sexuality and gender identity seem to center around prenatal hormone exposure. Although we think of these concepts as separate, we look for the same "cause" for them.

Despite most recent research about gender identity being "biological," there is evidence that there may also be social/environmental influences on gender identity. If we take a broad perspective, looking around we see that most societies have predefined roles for men and women and that most people stick to these relative boundaries, coming to understand themselves as the genders they were assigned at birth. In the 1970s, in a series of studies called the Baby X Experiments, researchers observed adults interacting with babies and noted that they communicated with and described the babies differently based on whether they were told the baby was a boy or a girl. Although it is difficult to prove a direct link, this research suggested that children are exposed to different social environments depending on their assigned gender, which has the potential to influence how they think about their gender identity. Evidence from studies of intersex people also suggests an influence of social environment on gender identity. Although some intersex people do transition to another gender from the one they were assigned at birth, most do not. That is, most continue to identify throughout life with their assigned gender, despite their genetics or prenatal hormone exposure.

In the nature versus nurture debate, no one side is ever going to "win." This is because biological and social forces are not separate, but rather work together in complicated ways. For example, we think of height as a genetically determined trait, but people from tall families raised without good nutrition end up shorter than the rest of their family members. Gender is even more complex. Without human language and culture, sex might exist, but not gender or gender identity. It is likely that each individual's gender identity is influenced by a number of different biological and social factors and that these may not be the same for everyone.

What does gender have to do with sexuality?

Gender and sexuality are often confused. Gender refers to our own sense of how we see ourselves, while sexuality is defined as the gender or genders of the people we are attracted to.

The acronym LGBTQ has come into common use to refer to a spectrum of people whose gender identities and sexualities fall outside the traditional cisgender heterosexual norm. While these groups can share some common experiences, they can also be quite distinct.

Gender identity refers to someone's inner sense of their own gender. Sexuality, on the other hand, involves a person's attractions. For example, a transgender woman assigned male at birth may identify as a woman (her gender identity) and may be attracted to any combination of people. She may label her sexuality as gay, straight, bisexual, queer, or something else.

Many would argue that together, each group within the LGBTQ spectrum is more politically powerful than it would be alone, but not everyone involved always sees their struggles as linked. A well-known example of a clash within these communities is the decision in 2008 by one of the largest LGBTQ organizations in the country, the Human Rights Campaign, to support a federal employment nondiscrimination bill that it felt was more likely to pass, but included protections based on sexual orientation and not gender identity. This led many transgender people and allies to point out the long history of trans discrimination and marginalization within gay and lesbian organizations.

Divisions in LGBTQ communities continue to exist today. In 2015, a change.org petition titled "Drop the T," suggested that transgender people should be dissociated from the larger LGBTQ spectrum. The author argued against supporting children in their identification as transgender and believed that trans people should not be allowed to use bathrooms matching their gender identities. Luckily, there are many organizations that attempt to connect rather than separate LGBTQ communities.

It is clear from the amount of confusion and conflict about gender identity and sexuality that the two concepts are linked in our minds. But why? What does our gender have to do with our sexual orientation?

Although we have moved forward somewhat in recent years, there are still numerous gendered expectations that guide our clothing, career choices, and everyday behaviors. Additionally, according to gender norms, men are expected to be attracted to women and women to men. Presumed heterosexuality, like other expectations, is built into our gender-related societal framework. Therefore, at the simplest level, minority sexualities (gay, lesbian, bisexual, and queer identities) challenge gender norms.

One of the most common types of bullying in schools involves homophobic language. Interestingly, however, this kind of bullying starts long before children reach an age when sexuality is emerging. Although children may call each other "fags" or other similar slurs, they are not responding to each other's sexuality, but instead to their peers' gendered presentations and behaviors ("feminine" behavior in boys and "masculine" behavior in girls). From a young age, children come to link sexuality and gender and to see those who transgress gender norms as being sexual minorities.

Some of the children who are teased because of their gender presentations do eventually fall along the LGBTQ spectrum, although many do not. Additionally, many adults who identify as lesbian, gay, or bisexual buck gender norms by wearing clothes or behaving in ways that challenge expectations for their genders. They may do this to signal their sexual orientation to others as part of queer culture, or they may simply be doing what feels right for them. There are also many lesbian, gay, and bisexual people for whom traditional gender roles feel right.

While LGBTQ culture has, in many ways, begun to embrace the idea of challenging gender norms of dress and behavior, straight culture has been slower to adopt these attitudes. Straight men, especially, often have a more difficult time than straight women stepping outside of gender expectations. This may have to do with the very rigid rules within which men are taught to exist in our society. Straight culture may have something to learn from queer culture, which has expanded our understandings of ourselves to include a wide variety of diverse ways of being.

2

THE HISTORY OF GENDER

How much do we know about early humans and gender?

Most people believe that gender equality is a product of modern times and that the gap between the sexes has decreased due to technology, advocacy, and contemporary social demands. We typically think about early humans as hunter-gatherers and assume that because of the natural capacity for male bodies to be physically stronger and more robust than female bodies, men were the ones primarily responsible for the livelihood and sustainability of a group. We have inferred that men were responsible for survival, and as such, they were likely leaders within groups, and the ones with the most power. We have assumed that women's roles as gatherers were less critical than their male counterparts and thus less valued. However, more and more evidence points to these assumptions being incomplete and sometimes wholly incorrect. Instead, we are learning that gender roles in early human communities were probably far more egalitarian than what we originally thought.

In the Paleolithic Age, or the Old Stone Age, before the advent of agriculture, it is likely that men and women shared most responsibilities. While select activities may have been dominated by one gender or another, the vast majority of survival-based activities were likely performed together. There is no evidence that stone-age tools were made only by men. Similarly, a 2012 study evaluating stone-age cave paintings revealed that the majority of handprints were from female bodies, disproving the widely held belief that men were primarily responsible for Paleolithic art. Children were likely raised communally, sometimes through the practice of alloparenting, where unrelated community adults cared for them.

The Neolithic, or New Stone Age, was a time wherein hunter-gatherer communities moved less, starting to farm for the first time. Prior to the New Stone Age, private property and material means were concepts not yet applicable because movement was constant. A family may have been continually moving, and thus even the concept of a semi-permanent household didn't yet exist. The move toward farming rather than hunting and gathering meant communities needed to remain in one place for months at a time. As such, ideas around the home and house became further defined. A new division of labor emerged wherein there was a delineation between duties that needed to be cared for within the house and home and outside. Often, gender determined who focused on what set of chores, creating clearer lines of segregation between what men and women did and how men and women would become valued.

One could argue that as communities became less nomadic and more invested in material value, gender roles became more distinct. A 2017 study examining gender inequities in the Central Plains of China during Eastern Zhou reveals a significant shift between the Neolithic Age and Bronze Age, just afterwards. By examining carbon and nitrogen isotopes in preserved bones, scientists were able to determine the types of plants and the amounts of animal products people ate in the decade preceding their deaths. Analysis revealed that in the Neolithic Age all individuals, regardless of gender, were consuming similar sources of nutrients. However, this changed in the Bronze Age, when new crops were introduced and domesticated animals became more prevalent. Men continued to live on traditional millet and animal products, while women relied mostly on wheat. Female skeletons became significantly shorter in the Bronze Age, presumed to be a result of childhood malnourishment—likely girls were the first to be deprived if there was a shortage of food. Researchers have understood this shift to be evidence that in the Bronze Age, men and women began eating differently and socializing separately. Further confirming this shift is the archaeological evidence uncovering Bronze Age inequities: Neolithic burial sites showed no clear evidence of gender inequalities, but in the Bronze Age, males began being buried with more riches.

Prior to the Bronze Age, males and females were dependent on each other for survival—both gender roles brought essential contributions to the group as a whole. Scholars attribute the decline in women's social status to the introduction of new crop plants and domesticated animals, devaluing the functional roles women had participated in, therefore decreasing their standing in their communities.

The emergence of agriculture also allowed humans, perhaps for the first time, to accumulate significant resources. In this system wherein resources and power could be more effectively hoarded by just a few, it became more beneficial for male parents to bond with their male kin, yielding gains that could be passed along through generations.

Many advantages were lost when societies switched from nomadic to farming lifestyles. In addition to the obvious benefits for women in egalitarian societies, researchers have found that equality between the sexes may have provided a more robust social network with wider choice of mates, providing a genetic advantage among early humans. This model suggests that for our early human ancestors, gender equality was highly beneficial, compromised only when wealth and production became prioritized with the emergence of agriculture.

If gender equity was so beneficial, why have we for so long assumed that early human societies were male dominated and run? Until recently, much of our data has been derived from studying living hunter-gatherer societies, such as 19th-century indigenous peoples of the Americas. While these groups may not have had the same structures as earlier nomadic societies, their organization is also likely to have been interpreted through a patriarchal lens by colonizers, obscuring any egalitarian arrangements present. All research is affected by the point of view of the researcher, and cultural scholarship is no exception.

What does evolutionary psychology have to say about gender?

In the 1859 book, *On the Origin of Species,* Charles Darwin argued that natural selection led to evolutionary changes by preserving the genes of those animals most fit to survive in their current climate. Those with the most useful features to propagate their genes passed them along to their offspring. This is how, the theory goes, we began to stand on two feet and the size and functionality of our brains diverged from common ancestors.

But not all humans alive today evolved in the same ways. There are ethnic groups where people tend to be shorter or taller, more or less hairy, darker or lighter skinned. Women and men from the same ethnic group also differ from each other. While there are tall strong women and short weak men, on average, men are taller and stronger. Women tend to have a higher percentage of body fat, likely to nourish growing fetuses, but there are also women with almost no body fat and men with much more. This, evolutionists argue, is a phenomenon called sexual dimorphism and is due to sexual selection, the theory that certain traits have evolved differently in men and women because they posed advantages to reproduction.

Evolutionary psychology theorizes that it is not just our physical characteristics that can evolve but also our psychological traits and behaviors. Human children, for example, are extremely good at learning languages, even if they have not been taught to them explicitly. It is likely that those people whose brains were better suited for learning language fared better than others in passing along their genes because they could communicate well.

When it comes to gender, evolutionary psychology is a hotbed of debate. Most scientists believe that men and women are, on average, physically different in some ways because of evolution. But are we also psychologically distinct due to evolution, or are our psychological differences a result of culture? Nonhuman animals show differences in behavior between the sexes, and they do not possess "culture" in the same way we do. In addition, certain traits appear to differ between men and women across many cultures. A well-known example is that men tend to be better at spatial tasks, and this finding holds steady when studies are done in many different places. However, there is also evidence that women who are primed to think that they are good at spatial tasks do better than those who are not and that women who play video games become better at spatial tasks, demonstrating that this trait may be influenced by learning and culture.

Where there is the most debate is around behaviors that are stereotypically male and female. Are men more aggressive, or women more nurturing, on average, because of evolution or because of social norms? Is it "natural" for men to want to sleep with as many women as possible, but for women to want to find steady partners, because these are the ways in which we can propagate our genes most successfully? Are men, by nature, more aggressive than women, because in the past they needed to be? Does this mean we should "allow boys to be boys?"

In addition to discussions about male versus female evolutionary traits, there are also questions about the evolutionary purpose of same-sex attractions and transgender identities. If the goal of attraction is to carry on our genes, isn't it counterintuitive to spend energy and resources on same-sex attractions? And where do transgender people fit in? How is it evolutionarily helpful to have an identity that does not match assigned sex?

Evolutionary psychologists have suggested a number of theories to explain homosexuality and bisexuality. One idea is that the same genes that tend to lead to same-sex attractions and decrease reproductive fitness in one person might boost the reproductive fitness of others in their family. Sickle cell trait is a good example of a gene that did not seem to have an evolutionary purpose until it was discovered that having two copies of the gene for sickle cell disease causes the disease itself, but having one copy helps to protect against the illness malaria. Using this reasoning, the genes that contribute to a man being gay might also be more likely to make his straight brothers or sisters more attractive to another sex. Arguments have also been made that genes involved in same-sex attraction can cause increased interest in nurturing siblings' children, therefore propagating a person's own genes indirectly.

Sexual practices and the evolutionary role of monogamy continues to be heavily debated. In 2018, Wednesday Martin published *Untrue: Why Nearly Everything We Believe About Women, Lust, and Infidelity Is Wrong and How the New Science Can Set Us Free*, attempting to debunk the belief that women have an instinctual drive to partner monogamously. Her argument highlights that, while data sets are difficult to validate, women may have just as much of a desire to have sex outside of their primary, presumed heterosexual and monogamous relationships as their male counterparts.

Gender diversity has existed throughout time, inclusive of individuals who have held identities that are incongruent or not fully inclusive of the genders associated with their sex assigned at birth. Evolutionary psychology actually supports subset diversity, as communities simply would not be able to sustain themselves over time if all members subscribed to one of only two narrowly defined roles. Individuals who fit into a third gender, or transgender identity, historically have been entrusted with critical community responsibilities that fell outside of the domains of men's and women's work, respectively.

It is unlikely that we will settle the debates about evolutionary psychology anytime soon. Human behavior is extremely complicated and is likely influenced by a combination of genes, social influences, and personal experiences. It is possible that there are some average differences between men and women related to our evolutionary past, but it is also evident that in modern society we are capable of making choices that contradict our genetic make-up. Whether or not we carry with us our evolutionary past, we can make decisions about our future. For example, even if it were true that men were naturally more aggressive than women, would we want to live in a society that tolerates violence? Evolution is dictated by survival of the fittest, but in today's world, shouldn't life be about more than just survival?

What is patriarchy?

Patriarchy is a social system in which men have power over women. Patriarchy may exist within households or within the greater society and government. The vast majority of societies today are patriarchal. Most patriarchal societies are also patrilineal, meaning that family names, titles, and property are passed down through the male side of the family.

Patriarchy is reinforced through sets of beliefs about men and women, specifically that men are superior to women. Rationalizations for patriarchy typically include arguments that men are biologically suited to lead, while women were created to serve or follow. Most justifications for male rule are based around contentions that men have a superior intellect and ability to rationally approach problems. Women are stereotyped as more emotionally driven and less intelligent than men. There are also religious rationales for patriarchy. In evangelical Christianity, for example, God is viewed as male, and followers believe that God created separate gender roles for men and women, with men expected to lead the household and the church.

Until 100 years ago the legal system in the United States prevented women from voting and participating in other civic activities such as sitting on juries. Women were considered the property of men, and married women could not own their own property. It was not until the mid-1970s that a man raping his wife was considered a crime in any state. To this day, in some states, there are separate laws for marital and nonmarital rape.

Have human societies always been patriarchal? Many anthropologists argue that hunter-gatherer groups were more egalitarian, both in general social structure and also with regard to gender roles, than the farming societies that developed more recently. There is some evidence that hunter-gatherer groups surviving into modern times do participate in more shared decision-making, including women's voices alongside those of men. The reasons for this difference are not completely clear. It has been suggested that a hunter-gatherer lifestyle meant that members of a band or tribe were highly dependent on one another. Often, survival hinged on food that could be gathered, usually by women. Hunting was less dependable and made up a smaller portion of calories. An argument has also been made that for a particular person or group of people to sustain power, they must have control over resources. In agricultural societies, land and food supply could be controlled, but in hunter-gatherer societies, there were not similar ways to accumulate wealth and therefore consolidate power.

Whatever the reasons for this shift from more egalitarian societies to our current, economically unjust and patriarchal societies, we have now been living under patriarchal rule for over 10,000 years. In that time, we have developed belief systems about male superiority that pervade almost every aspect of our lives.

What is matriarchy?

Greek mythology cites the Amazons, a tribe of women warriors, as an ancient matriarchy. Historical accounts document women leaders across the globe, some who fought alongside their male counterparts, others leading them in war. The modern day Wonder Woman, a Marvel comic, embodies the Amazon archetype, along other fictional and real-life characters that shape the way we understand gender, power, and leadership. But has a true matriarchal society ever existed?

Among the Mosuo (also called Na) people, a small group of about 40,000 living in China near Tibet, children are raised in large households by their mother's side of the family. Lineage is traced through the female side, and property is passed down the same way. Men live with their mothers' families, and couples have "walking marriages," where women choose their partners, walking to the man's house to spend time together and then walking home to stay with her own family, never living together. When a couple has a child, the child lives with the mother and her family. The father stays to live with his own mother's family and assists in raising his sisters' children.

The Mosuo are one of just a few existing societies considered matriarchal—where women rule. Others include the Minangkabau of Indonesia and the Akan of Ghana. However, a closer look indicates that while these societies are matrilineal (names and property are passed down through the female side of the family) and often matrilocal (spouses and children live with the female side of the family), they are not technically matriarchal, as their political systems are typically run by men. In certain groups, women have the power to remove male politicians they do not feel are acting appropriately, but these men are still replaced with other men. It is debatable whether there are any sincerely matriarchal societies in existence today.

It is also not clear whether there ever were any truly matriarchal societies in the past. Many of us share a collective myth that there was a time in history when matriarchal societies were the norm and that they were supplanted by patriarchal societies later. However, the historical record and anthropological evidence do not support this idea. Just as there are societies today that are matrilineal and matrilocal, these did exist in the past. Some worshipped goddesses and saw women as focal points in society. Native American groups such as the Hopi tribe have been described as matriarchal, because women ran the household and lineage was through the female side of the family. The Hopi are also said to have highly valued women's contributions to political decisions and to have included women in their governing bodies. Even this kind of a society falls short of matriarchal, though, and is likely better described as egalitarian, as women participated in government but did not rule it.

What did gender diversity look like in North America during colonization?

Individuals who today might consider themselves transgender have existed throughout time. Prior to colonization, many indigenous tribes in North America included a third gender category broadly referred to today as Two Spirit. Before the 1990s, Two Spirit individuals were commonly referred to as berdache by white, western society. The term is now considered offensive.

People in the Two Spirit tradition were vast and varied. According to the National Congress of American Indians, many served as mediators and were responsible for naming children, matching love partners, and serving as healers. Some were thought to be able to predict the future and bring good luck and peace. Many led puberty ceremonies. Varying by tribe, sex assigned at birth, and community needs at the time, the Two Spirit identity was diverse. So too were the words used to describe these individuals. The Blackfoot used the words Aakíí'skassi and Saahkómaapi'aakííkoan; the Cherokee, nudale asgaya and nudale agehya; the Lakota, Winkte and Bloka egla wa ke; and the Navajo, Nadleeh and nadle. At least 150 tribes are thought to have had names for Two Spirit individuals.

Colonization brought an abrupt and violent halt to many freedoms Two Spirit people had lived with for centuries. Historical accounts present the confusion early colonizers had upon meeting them. Missionaries were often cruel and inhumane and are reported to have fed Two Spirit people to the dogs, forced them into the clothing and hairstyles of a cisnormative and eurocentric standard, separated them from family, and altogether erased their histories. Whereas many Native Americans did not view gender as strictly male or female, european settlers came with the understanding that gender variation was unacceptable.

Early white settlers approached gender variance within their own communities similarly. Historical records show that a person named Thomas(ine) Hall, a Virginian servant who was likely intersex, was brought into court in 1629 after wearing varying male and female clothing. Hall was not given the choice to dress as they chose and was instead ordered by the Virginia court to wear both a man's breeches and a woman's apron and cap. Other court records confirm that gender expression was highly regulated in colonial society. In 1692, Mary Henly of Massachusetts was charged with illegally wearing men's clothing, "seeming to confound the course of nature."

How has the history of transgender people in the United States evolved since the country's founding?

During the initial century after the founding of the United States, gender diversity remained, for the large part, hidden from public view and was seen as aberrant. Some people did live lives we might consider transgender today. Assigned female at birth in 1829, Joseph Lobdell lived as a man for 60 years prior to his arrest and incarceration in an asylum after his birth-assigned gender was revealed. During the Civil War, in the 1860s, over 200 people assigned female at birth donned men's clothing and fought as male soldiers. Some were transgender and lived the rest of their lives as men, including Albert Cashier, an Irish-born immigrant who served in the Union Army.

In 1879, We'wha, a lhamana of the Zuni people, befriended anthropologist Matilda Coxe Stevenson in a Zuni pueblo in modern day New Mexico. The lhamana were third gender people who were male-assigned but lived performing more feminine roles in their communities. Per report, Stevenson did not realize that We'wha was not a cisgender woman until years after their friendship began. In 1886, We'wha visited Washington, DC, with Stevenson and several others and was introduced as "an Indian Princess" to 22nd and 24th U.S. President Grover Cleveland.

Late in the 19th century, gender-diverse individuals, not yet using the term *transgender*, began to organize. In 1895, a group of self-described "androgynes" in New York City organized a club called the Cercle Hermaphroditos, based on their wish "to unite for defense against the world's bitter persecution." Jenny June, one member of Cercle Hermaphroditos, wrote *The Autobiography of an Androgyne* (1918), *The Female Impersonators* (1922), and a third volume of memoirs (1921) that was not published until 2010. These texts continue to be regarded as rare, first-hand accounts of gender diversity in the early 20th century.

In 1917, Alan L. Hart, working with psychiatrist Joshua Gilbert, was the first trans man in the United States to undergo a hysterectomy and gonadectomy, steps taken to live his life as a man. Hart was a physician, radiologist, tuberculosis researcher, writer, and novelist. He pioneered the use of X-ray photography in tuberculosis detection and helped implement tuberculosis screening programs that saved thousands of lives.

Progress toward legitimacy and nonpathologization for transgender people has not been linear. Despite respect for Hart's medical expertise, his transition and leadership within the medical field did not bolster the overall rights of transgender people during his time. In 1945, trans woman Lucy

Hicks was tried in Ventura County for perjury and fraud for receiving spousal allotments from the military, as her dressing and presenting as a woman was considered masquerading.

Other people who would likely identify with the label *transgender* today lived in secret, at least when it was possible to do so. One notable example is American jazz musician Billy Tipton, who lived as a man in all aspects of his life from the 1940s until his death. His own son reportedly did not know of his past until Tipton's death. Tipton's sex assigned at birth was revealed across multiple media outlets in the days following his passing.

More well-known is the case of Christine Jorgensen, who in 1952 became the first widely publicized person to have undergone gender-affirming surgery. Jorgensen was a recent veteran, with access to travel to Europe for her surgeries. Despite fame, Jorgensen was denied a marriage license in 1959 when she attempted to marry a man. Her fiancé, Howard J. Knox, lost his job when his engagement to Christine became public knowledge.

Virginia Prince, who lived in San Francisco, developed a widespread correspondence network with transgender people throughout Europe and the United States in the 1940s, 1950s, and 1960s. Prince believed that the binary gender system harmed both men and women by keeping them from their full human potential, advocating for increased inclusivity and awareness across the fields of literature and social sciences. Prince worked closely with Alfred Kinsey, PhD, founded *Transvestia* magazine, and started the Society for the Second Self. Prince was one of the first people to use the term *transgender*, which, to her, referred to people like herself who lived full-time in their identified gender without the intention of having genital surgery. Prince was one of the first to distinguish between sex, sexuality, and gender identity. While her work was criticized for being too heavily reliant on gender norms and stereotypes, she continues to be regarded as one of the major pioneers in modern transgender history.

Other community-based organizations began to mobilize at this time. A wealthy transgender man named Reed Erickson founded the Erickson Educational Foundation in 1964, providing free information to transgender people, family members, and professionals. The Erickson Educational Foundation also funded the earliest symposia for professionals who worked with transgender people, a group that later emerged as the Harry Benjamin International Gender Dysphoria Association, which is today called the World Professional Association for Transgender Health. In the late 1960s in New York, Mario Martino founded the Labyrinth Foundation Counseling Service, and in 1972, Angela Douglas started TAO (Transsexual/Transvestite Action Organization). TAO grew into the first international transgender community organization, fighting against the pathologization of transgender people as mentally ill.

As trans people began organizing, so did their visible and vocal fight against transphobia. In 1959, the Cooper Do-nuts Riot in Los Angeles marked one of the first LGBTQ uprisings in the United States. Six years later in 1965, 150 people protested at Dewey's Coffee Shop in Philadelphia, which was known to have refused service to people in "nonconformist" clothing. In 1966, trans people in San Francisco's Tenderloin district protested at Compton's Cafeteria Riot, and in 1969 in New York, the Stonewall Riots marked one of the most violent and also mobilizing events in LGBTQ history.

Led by transgender women of color, the riots came in response to New York City police raiding the Stonewall Inn, a gay club located in Greenwich Village. "Solicitation of homosexual" relations was then illegal in New York City, and there was a criminal statute that allowed police to arrest people wearing less than three "gender-appropriate" articles of clothing. On June 28, 1969, armed with a warrant, police officers entered the club, roughed up patrons, and, finding illegally sold alcohol, arrested 13 people, including employees and people violating the state's gender-appropriate clothing statute. Provoked by violent arrests and lifetimes of social ostracization and discrimination, bar patrons and nearby locals fought back. Within minutes, a full-blown riot began. The protests, at times involving thousands of people, continued in the area for six days.

The Stonewall Riots, as well as the Cooper Do-nuts Riot and the Compton's Cafeteria Riot, served as a catalyst for the transgender rights movement in the United States and around the world, ultimately serving to progress legislation, services, and visibility. Following the riot at Compton's, a network of transgender social, psychological, and medical support services were established, culminating in 1968 with the creation of the National Transsexual Counseling Unit, the first peer-run support and advocacy organization in the world. Sylvia Rivera and Marsha P. Johnson, both present at the Stonewall Riots, went on to found the Street Transvestite Action Revolutionaries (STAR). The organization, established to provide housing for LGBTQ youth, opened the first "STAR House" in a parked trailer in Greenwich Village in the early 1970s. STAR House was the first LGBTQ youth shelter in North America, the first trans woman of color led organization in the United States, and the first trans sex worker labor organization.

By October 1979, the first National March on Washington for Lesbian and Gay Rights was held, drawing between 75,000 and 125,000 LGBTQ people and allies to demand equal civil rights and urge the passage of protective civil rights legislation. The march was organized by Phyllis Frye, who later became Texas's first openly transgender judge.

In the last two decades of the 20th century, transgender people became increasingly visible, by way of both increased scrutiny and responsive mobilization. In 1986, transgender activist Lou Sullivan founded the support group that grew into FTM International, the leading advocacy organization for transmasculine individuals. In 1991, the inaugural Southern Comfort Conference was held in Atlanta, Georgia. One of the first and largest of the time, the conference served to bring together trans community from across the United States to share resources and information, build community, and ignite advocacy.

In 1998, Karen Kopriva became the first American teacher to transition on the job. In 1999, Monica Helms created the modern day version of the transgender pride flag, which is now flown with pride at events all over the world. By the end of the 20th century, there was increased visibility of transgender people, and transgender legal rights became had become a focal point in modern American politics.

What has trans activism looked like in the 21st century?

Despite many of the advances made by trans people in the 20th century, the path toward visibility and acceptance has not been straightforward and has, in fact, harbored violent backlash. In 1988, trans woman and legendary ballroom performer, Venus Xtravaganza, was strangled to death in New York City. In 1993, 21-year-old Brandon Tina was raped and murdered in Falls City, Nebraska. Today, homicides of transgender people, particularly trans women of color, continue to escalate. The Human Rights Campaign tracked at least 21 murders of transgender people in 2015, followed by 23 in 2016 and 28 in 2017. Transgender communities memorialize these deaths every year on the Transgender Day of Remembrance, founded in 1998 by Gwendolyn Ann Smith to memorialize the murder of Rita Hester.

As the 21st century has moved forward, trans communities have pushed for change in every area of advocacy, including work to change social opinion, increase legal protections, and create policy to advance the rights of transgender people across the country.

In 2004, the first Trans March was held in San Francisco, and in 2005, transgender activist Pauline Park became the first openly transgender person chosen to be the grand marshal of the New York City Pride March. In 2006, Kim Coco Iwamoto was elected as a member of the Hawaii Board of Education. At the time, Iwamoto was the highest ranking openly transgender elected official in the United States. In 2008, Stu Rasmussen became the first openly transgender mayor in America, and in 2009, Diego Sanchez became the first openly transgender person to work on Capitol Hill, where he was a legislative assistant for Barney Frank. In 2010, Amanda Simpson became the first openly transgender U.S. presidential appointee, the same year Victoria Kolakowski became the first openly transgender judge. Sarah McBride was a speaker at the Democratic National Convention in July 2016, becoming the first openly transgender person to address a major party convention in American history. By 2017, transgender woman Danica Roem was elected to the Virginia House of Delegates. The same year, Tyler Titus, Phillipe Cunningham, and Andrea Jenkins, three transgender people, were elected into public office in the United States all on the same night.

Media and academia followed suit. Kye Allums became the first openly transgender athlete to play NCAA basketball in 2010. Chaz Bono made headlines when he came out in 2011 as a transgender man. In 2014, actor Laverne Cox became the first openly transgender person on the cover of *Time*, and, in 2015, Caitlyn Jenner came out in a television interview as a transgender woman. In 2014, *Transgender Studies Quarterly*, the first nonmedical academic journal devoted to transgender issues, began publication with two openly transgender coeditors, Susan Stryker and Paisley Currah. The same year, Mills College became the first single-sex college in the United States to adopt a policy explicitly welcoming openly transgender students.

Acceptance of transgender people in the United States has a long way to go. With record high rates of violence, transgender people today are regularly denied employment, housing, adequate medical care, and basic respect. Yet slowly, progress is being made. Policy has become largely more inclusive in the last decade. Cultural institutions like the Boy Scouts of America and the Girl Scouts of America, as well as major religious communities, have begun to institutionalize acceptance of transgender people. Progress has not been linear. In 2016, the Obama administration issued guidance that clarified Title IX protections for transgender students, which were later rescinded in early 2017 by the Trump administration. We have yet to see what trans activism will look like in the 2020s, but we can be sure that trans communities will continue to work toward a more just and respectful world.

3

GENDER AND BIOLOGY

What are the biological differences between male and female bodies?

Mass media blasts us with almost daily doses of ideas like "Men are from Mars; women are from Venus." We're told that men and women are physically, emotionally, and cognitively different from each other. But how much of this is true, and how much is stereotypes?

There are a few things we do know. On average, male bodies are taller than female bodies. On average, male bodies are also stronger than female bodies. It is important to keep in mind, though, that these are averages. Height and muscle mass vary widely across ethnicities, so that the average height of male bodies in some southeast Asian countries is around 5 feet 4 inches, while the average height of female bodies in some western and northern european countries is around 5 feet 6 inches. Neuroimaging studies that look at structural and functional brain differences have yielded conflicting results; to date, there are no decisive, reliable, category-defining differences between the male and female brain.

Physical development begins as a fetus grows in the uterus, and those people whose bodies will later appear female undergo a different set of changes from those whose bodies will later appear male. There is also a large group of people who are intersex, a term used to describe those who do not fit neatly into male or female boxes.

Embryology is the study of prenatal development and covers the events that happen from fertilization until birth. Human embryos, no matter their chromosomes, all begin with the same internal structures, and these remain until approximately six to eight weeks after conception. Internal and external genital development then occur along separate pathways, making it possible for a person to have internal organs that do not match those typical of others with similar external organs.

Internally, we each begin with two sets of ducts—mesonephric ducts and paramesonephric/Mullerian ducts. In those who will have female-typical internal organs, the Mullerian ducts develop into the uterus and upper portion of the vagina. Those with a gene called sex-determining region Y (SRY) gene, typically found on the Y chromosome, develop testes, which then secrete testosterone and anti-Mullerian hormone (AMH). One of the functions of testosterone is to further the development of the mesonephric ducts, which become the epididymis and vas deferens, structures that later carry sperm from the testes out of the body. AMH signals the Mullerian ducts to recede.

While the SRY gene is usually found on the Y chromosome, there are also people who have a Y chromosome with no SRY gene or have an SRY gene on another chromosome. This can lead to differences from typical male or female development. There are also some people whose AMH receptors do not respond to AMH. They may have male-typical internal and external organs but may also retain their Mullerian ducts and have a uterus and a portion of a vagina.

Externally, all fetuses begin with similar structures, called the urogenital groove and labioscrotal folds. Around 8 to 12 weeks into development, depending on the hormones secreted, these become either a clitoris, vagina, and labia or a penis and scrotum.

In some people with XX chromosomes, there may be more than the typical amount of testosterone, leading to the development of external genitals that appear somewhat between average male and female genitals. There are also some people with XY chromosomes whose bodies do not produce the enzyme that converts testosterone to dihydrotestosterone (DHT), which typically "masculinizes" the genitals. These people, with five alpha-reductase deficiency, may have female-appearing external genitals as children, but as they approach puberty, their bodies change to appear more masculine.

During puberty, our bodies differentiate from each other in a variety of ways. Those with XX chromosomes typically begin puberty a year or two earlier, around age 10 or 11, and their bodies respond to a number of different hormones, including testosterone, estrogen, and progesterone. Those with XY chromosomes also have all of these hormones in their bodies, but typically have a higher testosterone surge.

Early signs of puberty for many people, regardless of sex, are oily skin and armpit/pubic hair growth. Those who develop female "secondary sex" characteristics usually grow breasts, begin to menstruate, and their bodies have a higher fat-to-muscle ratio. Those who develop male secondary sex characteristics grow more body hair, have deeper voices, and have a lower fat-to-muscle ratio.

In popular culture, estrogen and testosterone are viewed as "sex hormones," associated with either femininity or masculinity, and linked to stereotypical female and male behaviors. In fact, everyone, regardless of sex or gender, has some level of both of these hormones in their bodies, and each hormone has multiple functions, some of which have nothing to do with typically feminine or masculine bodies or activities. Estrogen, for example, is produced, among other places, in the sertoli cells of the testes and is important for bone health. Testosterone is produced by the ovaries and adrenal glands in addition to the testes and contributes to sex drive, muscle mass, and red blood cell production in everyone.

We often talk about estrogen and testosterone as causing biological differences in the way men and women think and behave. However, there is not clear evidence that these hormones lead to the stereotypical differences we often associate with them. It would be difficult to design an experiment to measure hormones' effects on our thoughts and behaviors because we would need to randomly assign certain people, starting before birth, to receive specific hormones, and others to receive different hormones. We would also have to control for social factors that likely influence the way men and women think and behave. In other words, we would have to raise those in the study in completely gender-neutral ways.

For now, what we do know is that typical male and female bodies differ from each other in predictable ways, such as height, strength, and genital/reproductive capacity, but that there are many exceptions, and many people who do not fit neatly into these two boxes.

How similar or different are male and female brains?

Society would have us believe that women and men think so differently that we are almost separate species. Men are seen as rational, while women are supposed to be more emotional. We're told that women are caring and men are practical and that men have a better sense of direction and women have a better understanding of interpersonal interactions. Where do these stereotypes come from? Do they hold any weight? If not, why do we cling to them?

One consistent difference between brains of people of different sexes is their size. Similarly to male and female bodies, male brains tend to be bigger than female brains. In the 19th century, male scientists used this fact to argue that women were not as smart as men. However, we now know that this difference in the size of our brains is simply reflective of our height and weight differences. In other words, brain size is proportional to body size. While women's brains are, on average, about 10% smaller than men's brains, they are more condensed, with more gray matter (the part of the brain that contains cell bodies).

What about IQ? IQ stands for intelligence quotient. IQ tests were first designed in the early 20th century as standardized methods of comparing intelligence across populations. There is significant controversy over the accuracy and utility of intelligence tests. One criticism is that there are many types of intelligence, and IQ tests generally cover only a few areas. Verbal and math skills, for example, are commonly tested, but creativity and social intelligence are not. Another area of contention regarding IQ tests is their biases. Black Americans are not less intelligent than white Americans, but they do, on average, have lower IQ test scores, reflecting differences in socialization and racial biases in testing.

There has also been a strange phenomenon over the last century in which IQ scores for younger generations are rising compared to older generations, and scoring has had to be adjusted to account for these changes. There are many proposed explanations for rising IQs, including better schooling, test familiarity, and nutritional status, but none have to do with changes in genetics, meaning that IQ is likely not a measure of inborn intelligence and is affected by social and educational factors.

When it comes to sex and IQ, there is also significant controversy. In 2005, Lawrence Summers, president of Harvard University, gave a speech in which he stated that men were more likely to be geniuses than women and that one explanation for this could be genetic differences in intelligence. He did not take into account differences in socialization. In countries with higher levels of gender equality, for example, gender gaps in math scores have disappeared. Some researchers argue that, while men and women may not differ in their average scores on standardized tests, men's scores tend to fall more toward the two ends of the bell curve. That is, they are more likely to have extremely low or extremely high scores than women are. This is an area of ongoing investigation.

In addition to broad intelligence tests, researchers have also investigated possible differences between men and women on specific cognitive tasks. Men, for example, do better, on average, on tests of spatial ability. However, there is not clear evidence that this difference is genetic. Women's scores on spatial tasks improve when they practice by playing video games or when they are primed to believe that women are good at these tasks, suggesting that male advantage may be at least partly due to socialization.

Can we tell anything about male and female brains by looking at them? A number of studies have found average differences between men and women in the size or shape of particular brain regions. There does seem to be an area of the hippocampus (memory center) that is, on average, larger in men than women. However, most studies of particular brain regions have been difficult to replicate, meaning that they did not show similar results when different sets of researchers looked at the same thing.

Men are more likely than women to be left-handed, and handedness can sometimes correlate with lateralization of the brain—that is, the side of the brain that we use to perform certain cognitive tasks, such as language. This suggests that men and women may, on average, have different lateralization of their brains, although most people's brains are set up similarly.

In neurological research, each study tends to be small, but meta-analyses (review papers that gather together evidence from many smaller studies) seem to show that there are average differences between men and women in the size of grey matter areas (cell bodies) and white matter areas (cell axons) in certain parts of the brain. The corpus callosum, for instance, connects the two hemispheres of the brain and appears to be larger in women than men. However, it is not clear what these differences mean and whether they are genetic or the result of socialization.

If we take a step back and look at the evidence related to sex differences in our brains, it becomes clear that men and women are much more similar than they are different. So what makes it so appealing to view us as having wildly different brains? Perhaps if we were to accept that we are, on average, similar in intelligence and emotionality, it would mean that we would have to reconsider our stereotypes about men and women.

What are sex chromosomes?

Until the early 20th century, there was no common understanding of how a person's sex was determined. Scientists did not know why some people's bodies appeared male and others female. It wasn't until 1905 that two scientists, one of whom was a woman (Nettie Stevens), separately were credited with discovering human sex chromosomes.

What exactly are sex chromosomes? To begin, it's important to talk about what a chromosome is. Deoxyribonucleic acid (DNA) is the genetic material that almost all living organisms use to store and pass on information. DNA is typically folded into a more condensed structure called a chromosome, which allows a large amount of DNA to fit into a small area.

Most bacteria have a single circular chromosome because they have smaller amounts of DNA than animals. Animals typically have a number of paired chromosomes. For example, dogs have 39 pairs of chromosomes and cats have 19 pairs. A greater number of chromosomes does not always signify more genes. Although humans have only 23 pairs of chromosomes, compared to dogs' 39 chromosomes, both humans and dogs have around the same number of genes (20,000).

The majority of humans have 23 pairs of chromosomes, divided into autosomal chromosomes (numbered 1–22) and sex chromosomes (X and Y). The sex chromosomes are called X and Y because of their shapes. When an embryo is created from an egg and sperm meeting, the egg typically contributes an X chromosome and the sperm either an X or Y chromosome.

Most people with female bodies have two X chromosomes (XX), and most people with male bodies have one X and one Y chromosome (XY).

Some people have a slight variation in their chromosomes. For example, those with Down syndrome have an extra copy of chromosome 21 (trisomy 21). When the variation is in one of the sex chromosomes (X or Y), the person is considered intersex. One type of variation involving the sex chromosomes is Klinefelter syndrome, where a person has two X chromosomes and a Y chromosome (XXY). People with Klinefelter syndrome appear male but may have less body hair, some breast growth, and infertility issues.

While sex chromosomes take their name from their contribution to sex determination, they actually play a number of roles in other parts of our lives. The X chromosome contains about 1,000 genes, and most have nothing to do with sex. They represent diverse functions spanning from color vision to blood clotting to hair growth. The Y chromosome, on the other hand, contains only about 200 genes and is much smaller than the X chromosome. Because not everyone has a Y chromosome, it does not contain essential genes. However, it is involved in immune system regulation, cancer growth, and heart disease risk.

The functions of the sex chromosomes are not limited to sex determination, and they are also not the only chromosomes that affect this process. Some autosomal chromosomes contribute to sex determination. While the SRY gene on the Y chromosome triggers the cascade that leads to male sex organs, the SOX9 gene on chromosome 17 is also essential, and without it, the body develops as female.

Some people wonder whether there are ways to change the likelihood that certain sperm (carrying either an X or Y chromosome) meet an egg for fertilization. Urban legends suggest that couples can have sex at particular times during the month to increase the likelihood of having an XX or XY baby, with some suggesting that, because Y chromosomes are smaller, sperm carrying them can move faster. Studies using computer-assisted sperm analysis have failed to show any differences in motility of sperm carrying X or Y chromosomes. However, there is evidence that sperm sorting, performed in a lab, may have an accuracy rate of around 75% to 85%. This sorting is typically done by treating sperm with a fluorescent dye. Because X chromosomes have more DNA than Y chromosomes, sperm can be separated based on how bright the dye appears.

Interestingly, while human sex differentiation largely rests on chromosomal variance, not all animals' sex is determined similarly. The sex of alligators, for example, depends on the temperature of their nest.

What does testosterone do?

The word *testosterone* evokes images of weightlifters, warriors, and other hypermasculinized figures. We typically associate testosterone with men and masculinity. However, both testosterone and estrogen, which is often considered a female hormone, are present in all of our bodies. These hormones are close enough in structure that a single enzyme, called aromatase, can convert testosterone to estrogen. In addition, each hormone has multiple functions in our bodies outside of sexuality and reproduction. Testosterone, for example, increases the production of red blood cells and contributes to bone density.

Testosterone plays different roles over the course of the lifespan. As early as fetal development, hormones can affect the appearance of our bodies. Those with XY chromosomes typically have a gene on the Y chromosome that signals the growing fetus to create testicles. The testicles then produce testosterone, which leads to the development of external male genitals (penis and scrotum). Some argue that testosterone exposure as a fetus can lead not only to masculinization of the body but also the brain.

In a famous experiment in 1959, researchers found that by castrating male guinea pigs or giving testosterone shots to female guinea pigs early in life, they could alter their sexual behavior. Female guinea pigs given testosterone attempted to mount other guinea pigs, a behavior typical of males, and castrated male guinea pigs positioned themselves to be mounted, a behavior typical of females. It is difficult to say what this change in behavior means exactly. Does it have to do with sexuality, gender, or something else? We have no way to know. What we do know is that early hormone exposure appears to be capable of affecting our brains and not just our bodies.

Despite exaggerated claims to the contrary, research demonstrates that men's and women's brains are much more similar than they are different. In fact, it is often impossible to tell the difference between a male and a female brain by inspecting it or by viewing the results of magnetic resonance images or computer tomography scans. However, many researchers believe that there are subtle, average differences between brains of different genders. On average, for example, men tend to perform better on spatial tasks than women do, although one particular woman may perform better than one particular man. Some researchers argue that this average difference is due to early hormone exposure. Autism, described as a disorder by some and a difference in cognition by others, is much more prevalent in males (although it may be underdiagnosed in females). It has been suggested that the increased rate of autism in males is also related to early hormone exposure.

Testosterone's first role is to act on the developing fetus to guide its physical development and likely also its cognitive development in ways that are still unclear at this time. The next stage at which testosterone acts is puberty. While each individual begins puberty at a slightly different time and proceeds at their own rate, there are some generalities that apply to most people.

In those with testes, testosterone is produced by the testes and also by the adrenal glands, which are found just above the kidneys and manufacture a variety of hormones. The adrenals are best known for producing the stress hormone cortisol.

It may be surprising to learn that in those with ovaries, testosterone is produced in the adrenal glands, but also in the ovaries, alongside estrogen. Adult testosterone levels in those with testes typically range from 250 to 900 ng/dL and in those with ovaries, from 15 to 70 ng/dL.

During puberty, testosterone has a number of effects on different areas of the body, including muscle mass, facial structure, body and facial hair, and phallus size. Trans men who take testosterone often experience changes similar to cisgender boys, although effects may be slightly different due to age. One of the first changes most people notice with an increase in testosterone level is acne. Testosterone stimulates oil glands, leading to clogged follicles. For most people, acne is manageable, but for some, it can be serious and leave scarring.

Testosterone also leads to body and facial hair growth, causing thinner, finer hair to develop into coarser, thicker hair. A person's ability to grow a beard is largely dependent on genetics. For many people going through puberty, facial and body hair growth are wanted changes, but for others, they are not. Hair changes induced by testosterone are permanent. Even with a drastic reduction of testosterone, be it from age or medical intervention, most people whose hormones played a part in their hair growth will keep these changes throughout their lifetimes. Electrolysis can be effective in removing unwanted typical "male-pattern" facial and body hair, a process some transgender women pursue if accessible.

Receding hairlines are also linked to testosterone. Specifically, testosterone is converted to another hormone called DHT, which can lead to hair loss. The way that each person's hair responds to DHT has to do with their genetics. Trans men who start testosterone in their late 20s or 30s can sometimes begin to have balding soon into their transition, at the age when others in their family began to go bald. In addition, this kind of hair loss is irreversible, so trans women who have already begun balding prior to physically transitioning cannot regain hair that has been lost.

Another irreversible effect of testosterone is deepening of the voice. Testosterone leads to thickening and lengthening of the vocal cords, which do not return to their original state if testosterone is stopped. The Adam's apple common in cisgender men is cartilage growth induced by testosterone to protect the vocal cords. Trans women whose voices have already dropped often utilize speech therapy to adjust their pitch. Trans men who take testosterone undergo a deepening of their voices but may also use voice training to sound more masculine, as gendered speech does not depend solely on pitch, but also involves aspects such as word choice, intonation, and volume.

While there are clear physical changes with testosterone during puberty or when taken as an adult, there is debate about testosterone's effects on cognition and behavior. There does seem to be clear evidence that testosterone can increase sex drive, so much so that it is prescribed to cisgender women for low libido. However, testosterone itself does not wholly determine sex drive. Cisgender men with depression who have testosterone levels within normal range can still have very low sex drives due to their mental health symptoms. Humans are complicated, and so is their expression of their sexuality.

Another complex human behavior is aggression, which is often linked in our minds to testosterone. However, studies attempting to link testosterone and aggression have produced extremely varied results. What we do know is that it is not simply a matter of increased testosterone levels leading to increased aggression. Like sexuality, aggression is situation-dependent and revolves largely around social circumstances and interactions.

Recently, magazines and TV channels directed at cisgender men have begun to advertise treatment for "low T," or low testosterone, also known as hypogonadism. These commercials suggest that vague symptoms like fatigue and weight gain may be related to insufficient testosterone levels. While there are men who experience this phenomenon, the majority of cases involving these kinds of symptoms turn out to be something else, such as depression. For those with diagnosed hypogonadism, testosterone supplementation can be beneficial. However, there are risks to taking extra testosterone if it is not needed, including possible heart problems. While there is a natural decrease in testosterone levels with aging, most men do not report significant issues and do not require supplementation.

What does estrogen do?

Like testosterone, estrogen has been dubbed a "sex hormone." It is thought of as being relevant only for female bodies but is present in everyone and has multiple functions outside of sexuality and reproduction. Among other things, estrogen contributes to bone health and also increases "good" (high-density lipoprotein) cholesterol and decreases "bad" (low-density lipoprotein) cholesterol.

Unlike the testes, which produce testosterone in utero, the ovaries do not manufacture estrogen while the fetus is developing. Instead, they remain quiet until puberty. After puberty, in those with ovaries, estrogen is mainly produced there. However, in both those with and without ovaries, estrogen is also synthesized in many other areas of the body, including the liver, pancreas, bone, and adrenal glands. Additionally, the body can convert testosterone to estrogen through the actions of the enzyme aromatase.

Average estrogen levels in cisgender men are typically between 10 to 40 pg/mL. Levels in cisgender women vary significantly throughout the menstrual cycle, but range from approximately 25 to 450 pg/mL. After menopause, levels can drop to less than 40 pg/mL.

Estrogen is the primary hormone driving pubertal changes in cisgender women. Transgender women may also take estrogen, sometimes with a testosterone blocker (often spironolactone). As estrogen levels rise, a number of changes occur, including breast development and body fat redistribution to a more traditionally feminine body shape.

Trans women who take estrogen as adults after undergoing male puberty typically experience similar changes in body fat and breast growth. They may also notice decreased muscle mass and softer, less oily skin. However, estrogen does not reverse all of the effects that testosterone had earlier in life. Irreversible changes from testosterone include deepening of the voice, body and facial hair growth, and changes in facial structure. Estrogen also does not cause regrowth of hair that was lost to male pattern baldness.

Menopause is often a time when the topic of estrogen comes up. In their late 40s or early 50s, people with ovaries may begin to experience changes in the patterns of their periods, hot flashes, mood swings, vaginal dryness, and other symptoms. The eggs in the ovaries age and decrease in number, which leads to fluctuations in hormone levels as the feedback loops between the ovaries and the body's hormones are thrown out of balance. Menopause is not a straightforward process of gradual decreases in estrogen levels. In fact, for a few years, estrogen levels may intermittently be 20% to 30% higher than they were previously. Because estrogen contributes to bone health, once through menopause, there is a higher risk of developing osteoporosis and bone fractures. Heart disease also increases, although there is debate about the role of estrogen in this process.

Estrogen can sometimes come up in conversation related to breast cancer. There are certain types of breast cancer that are considered estrogen receptor–positive. In these cases, the cancer cells have estrogen receptors that allow them to grow in response to estrogen. Treatment for estrogen receptor–positive breast cancers may include a selective estrogen receptor modulator such as tamoxifen.

Why do women live longer than men?

On average, across the world, women live about three years longer than men. Somewhat counterintuitively, women outlive men despite social inequality, unequal pay, higher rates of abuse, and increased levels of depression and anxiety. This gender gap in longevity varies from less than 1 year in India, to 6 years in the United States, to a staggering 12 years in Russia. However, it is consistent and well documented. And it is not completely clear what drives it.

The fact that the gap is wider in certain countries suggests that there are social and cultural influences involved. In Russia, for instance, alcohol and tobacco use are thought to have a significant impact on men's health and longevity. Throughout the world, substance abuse is more common in men. Men are also less likely to be adherent with medical treatments and more likely to die from accidents and homicide. It is more socially acceptable in most cultures for men to take risks; risk-taking is built into gender roles and ideas about masculinity. Although women are more likely to attempt suicide, men are more likely to die by suicide, as they tend to choose more lethal means, such as firearms.

Research has also focused on possible biological influences on the gender gap in longevity. Those with XY chromosomes appear to have a disadvantage from the very beginning. XY infants have an over 25% greater chance of dying in the first year of life than XX infants. Some have argued that the structure of our chromosomes affects our survival. While most women have two copies of the X chromosome, which is large (approximately 1,000 identified genes) and performs many different functions important to the body, most men have one X chromosome and one Y chromosome, a smaller piece of DNA with only around 200 identified genes. Having two copies of the X chromosome may be beneficial for women because there is a "backup" copy available, providing extra protection against harmful mutations.

Scientists have also suggested that the hormones estrogen and testosterone may affect longevity. Testosterone has been shown to increase "bad" cholesterol and decrease "good" cholesterol, while estrogen seems to do the opposite. Men are more likely to die early from heart disease, which is linked to both biological and lifestyle factors. There is evidence that eunuchs—those people who, in certain cultures, had their testicles removed—lived longer than other men. Animal studies also appear to show that female animals that have their ovaries removed do not live as long as other female animals.

Finally, there are simpler explanations that have been proposed for the difference between male and female longevity. Taller, bigger people (who more often tend to be men) have more cells in their bodies, making them more susceptible to cell mutations, which can lead to cancer. Bigger people also use more energy and burn more calories, putting more wear and tear on their bodies.

A recent global study of 101 mammal species found that, on average, females lived 18.6% longer than males from the same species, a difference much larger than that between human men and women, which hovers around 8%. Researchers propose that this disparity is likely due to environmental conditions, sex-specific genetic variations, and the way in which different sexes utilize their resources. For instance, in the case of bighorn sheep, when natural resources were consistently available, there was little difference in lifespan between males and females. However, in areas with extreme weather conditions, males were found to live much shorter lives, likely because they depleted their resources through sexual competition and the growth of larger body masses.

Whatever the reason for the gap in life expectancy for women and men, it is clear from the wide variation across the world that lifestyle matters. Which means that all of us, no matter our sex or gender, have the possibility of living longer if we lead healthier lives.

What do we know about sex and gender in nonhuman species?

Have you ever gone to a pond and watched the ducks? What is most striking about them? The mallard is one of the most common species of duck in the world and is found throughout Asia, Europe, and North America. Looking out onto a pond of mallards, half of the ducks (the males) have beautiful green heads, while the other half (the females) are various shades of dull brown.

What about us? How different are men's and women's bodies? Men are generally taller, stronger, and more hairy. Women have larger chests and men have larger external genitals. But there are men who are frequently mistaken for women and vice versa.

The amount of difference between males and females in a species is called sexual dimorphism, and many species are much more sexually dimorphic than we are. One of the best-known examples of sexual dimorphism is the peacock and peahen. While the peahen is certainly beautiful, the peacock's elaborate plumage can be almost breathtaking.

Other examples of sexual dimorphism include species in which males and females differ significantly in size. Mandrills, primates living in tropical rainforests in Africa, can be distinguished by both their size and coloring. Males typically weigh between 40 and 80 pounds, and females, between 20 and 30 pounds. On the most extreme end, the female triplewart seadevil fish is approximately 25 times as big as the male, measuring 12 inches, while the male is less than half an inch.

Sexual dimorphism can develop due to a number of factors. Females of a species may be larger because this allows them to produce more offspring and provide better parental care. In species where there is strong competition by males for females, males often develop characteristics that distinguish themselves as attractive or healthy and, over time through evolution, can come to look very different from females of the same species.

There are also species in which the two sexes are more similar than humans in certain ways. The female spotted hyena, for example, has a clitoris that is as long as the male's penis. For intercourse to occur, the male must enter the female through her retracted clitoris. Female spotted hyenas also urinate and give birth through this same organ.

In some species, there is no division between male and female, and instead, all members of the species have both male and female genitals. Banana slugs, named for their bright yellow color, are hermaphroditic. They each have a penis on their heads, and when mating they simultaneously enter each other to deliver sperm, then find a hiding spot and lay their eggs.

Some animals do have a sex, but this sex can change over the course of their lifetimes. In the female-dominated world of the clownfish, each clownfish is born with the ability to become male or female and starts off male. Schools of all-male clownfish are run by a female leader, along with her submissive male partner. When the female lays eggs, her male partner guards them until they hatch. When she dies, her male partner physically transforms into female and takes over as leader of the school. Another type of fish, called wrasses, experiences almost the opposite. When the male leader of a school dies, a female takes over, growing male sex organs.

Sex changes in fish species reveal that there is a social hierarchy and specific sex roles. Humans are no different. We have expectations based on sex that determine how someone should behave, what kinds of roles they can play in families and work worlds, and who they should have sex with.

When humans act in stereotypical ways, such as men being promiscuous, we often point to nonhuman species, arguing that we can't be blamed for behaviors that are part of our "nature." Similarly, we make judgments about behaviors we find sinful or offensive, such as homosexuality, and use animals' lack of that behavior to contend that it is "unnatural." However, a quick survey of the animal kingdom reveals that animals have diverse and interesting gender roles and sexual appetites.

Although males are typically thought of as more aggressive and therefore better at hunting, female lions (lionesses) catch more prey than male lions. While we think of it as "normal" for animal males to physically fight over females, certain species of female antelopes lock horns in battle over their desired males.

Male members of certain animal groups have also been known to take on what we think of as traditionally female activities. Male penguins stay at home and sit on their eggs while female penguins hunt for fish to feed their offspring. During this time, the male penguins fast, and if their chicks are born before the mother returns, the father's body produces and regurgitates a curdlike substance to feed the baby. In some species, fathers are even the ones to become pregnant. When seahorse males mate with seahorse females, the females insert their eggs through a tube into the males' abdominal pouches, where sperm and egg meet. The males then carry the embryos until they are ready to enter the world and care for themselves.

When it comes to sexuality, it would be very difficult for us to take our cues from animals. While there are certainly many species where heterosexuality is the norm, there are also many that have unique ways of interacting sexually. Bonobos, known for their extremely sexual lives, are bisexual, and most of their sex is not for reproduction. Homosexuality is found in numerous species, including Japanese macaques and bottlenose dolphins. A small percentage of domestic sheep have been found to prefer homosexual relationships over heterosexual relationships, even when members of the other sex are available for mating.

Sexuality is complicated and does not simply reflect the sex of an animal's partners. Another dimension of sexuality is the propensity to mate with one partner (monogamy) versus multiple partners (polygamy). Monogamous animals include beavers, otters, wolves, barn owls, bald eagles, and more. Polygamy is much more common than monogamy. In most polygamous species, the males are more promiscuous than the females, but this is not always true. Honey bees and mole rats both have social systems where a female queen has liaisons with males, but most of the rest of the colony is made up of workers who do not participate in sexual activity at all. Like other animals, humans are unique and possess our own distinctive ways of interacting sexually.

4

GENDER, MEDICINE, AND PSYCHOLOGY

What does gender development in children look like?

Infants as young as three to four months old can tell the difference between male and female faces, and by six months they can match these faces to male and female voices. Before a child reaches their first birthday, they are often able to associate faces of men and women with gendered objects like jewelry and tools, suggesting that early on, children are able to categorize by way of stereotypes.

Between one and three years of age, children rapidly develop language and understanding of gender. Using gender labels in speech usually manifests between 18 and 24 months. Toddlers who know and use gender labels are generally more likely to show a preference toward gender-stereotypical play with toys. In general, toddlers spend a longer time investigating gender incongruent than gender congruent behavior, indicating that they understand the difference. Despite their burgeoning ability to separate what are considered male and female attributes and behaviors, children at this age maintain significant flexibility around gender. A three-year-old may inquire if an adult was a boy or girl growing up. They may also be assigned male at birth and state that they would like to grow up to be a mother or female at birth and want to become a father, not knowing that most adults would not consider this possible.

Most four-year-olds will be able to tell you their gender and the genders of their peers. They will also be able to tell you that the girls in their classes are going to grow up to be women and the boys, men. But probe a little further and you may be surprised by the concepts they don't yet seem to understand. Try posing this question: "Uncle Jim has short hair. If Uncle Jim were to grow his hair long, would he be a man or a woman?" Many four-year-olds would tell you that if Uncle Jim grew his hair long, he would be a woman.

In the 1960s, researchers studying young children began to theorize about how children learned about gender. They fell into two main "camps," the first being those who believed that children's understanding of gender came from their environments (social learning theory) and the second being those who thought that children's brain development determined their ability to understand gender (cognitive developmental theory). Both groups of researchers had reasons to believe that their ways of thinking were correct.

According to social learning theory, children learn about how gender functions in the world through modeling of gendered behaviors, experiencing rewards and punishments for their own gendered behaviors, and being directly taught ideas such as boys being smarter and girls kinder. There is significant evidence for social learning theory if we compare men's and women's behaviors around the world. Those who grow up in areas where gender roles are more rigid tend to see themselves and others as more strictly defined by their gender roles, while those who learn about gender in more open environments allow themselves and their peers more freedom of expression.

Cognitive developmental theorists also saw their theories play out in observations of children. They were able to predict that children of certain ages would grasp concepts that they had been unable to understand at earlier ages. One well-known test is to show a child two beakers, each containing the same amount of water. The researcher then pours the water from each of these beakers into a separate glass—one tall and thin, and one short and wide. When asked to compare the amount of water in the two glasses, most children under seven will say that the tall thin glass has more water, despite having just watched identical amounts of water being poured into each glass. Older children understand the concept of "conservation" and correctly assert that the two glasses have the same amount of water. Cognitive developmental theorists believe this is because children's brains develop in a predictable pattern over time.

There is evidence that cognitive development affects children's understanding of gender. Psychologist Lawrence Kohlberg posited that there are three main cognitive stages of gender development in children. In the gender labeling stage (ages two to three), children begin to understand what gender is and start to be able to identify their own gender and the genders of the people around them. In the gender stability stage (ages three to five), children learn that gender is a stable trait in most people and that boys generally grow up to be men and girls to be women. Finally, in the gender constancy stage (ages five to seven), children realize that gender does not change with shifts in clothing or behavior and that a man remains a man if he puts on a dress or grows his hair long.

Sandra Bem's gender schema theory is another way of thinking about how children's cognitive processes affect their ability to understand gender. A schema is a framework we use to understand a concept. For example, we may have a schema for a book, in which we have learned that a book is something that we can hold in our hand, has pages we can turn, and has words on the pages. If we come across a new type of book, such as one that is too big to hold in a hand, has pictures but no words, or does not have pages we can turn (e.g., an e-book), we have to revisit our schema for a book and decide if the new book changes our schema of books in general or should be let into the category as an exception.

Schema theory, as it is applied to gender, suggests that children develop schemas for "male" and "female" and that they test new information against these schemas to determine whether they should change their schemas or make exceptions. There is evidence that children pay more attention to the schemas that apply to their own gender, likely because they are trying to learn as much as possible about how they are supposed to interact in the world.

There are numerous criticisms of both social learning and cognitive developmental theories of gender. Critics of social learning theory argue that our brains are hardwired to understand concepts at certain developmental stages and that children do not learn about gender purely through social forces such as modeling and reinforcement. If they did, we could teach very young children to understand gender as adults understand it.

Those who question cognitive developmental theories of gender point out that gender is a largely social phenomenon and that without social influences, humans would have no concept of gender. Some have questioned whether the stages of cognitive development around gender actually represent what they claim to represent. For example, in the gender constancy stage, when children are thought to be learning that gender is a constant trait regardless of clothing or behavior, some social theorists argue that what children are really learning is that genitals (representative of our sex) are constant and that our society equates genitals with gender. If this is the case, then children may innately be understanding of transgender identity in a way that adults have unlearned through societal teaching.

An unspoken assumption in much of the gender development work to date is the alignment of bodily sex with gender identity. While there is an expected degree of gender creativity in childhood play, children who do not develop in accordance with societal expectations have historically been pathologized as not having achieved a critical developmental milestone. Moreover, models to explain gender development in young people have relied heavily on a binary approach to gender, without room for identifying in other ways.

Only recently has transgender identity begun to be considered a valid developmental trajectory. Historically, transgender children have presented with heightened levels of depression and anxiety, among other signifiers of distress, including poor school performance and poor social integration. However, recent studies have shown that transgender children match their cisgender peer groups when raised in affirmed settings. Poor functioning has been reconsidered not to be a result of "not doing gender right," but rather a psychological response to the experience of minority stress, hostility, and isolation.

Notably, research demonstrates that transgender children and their siblings view gender as more flexible than other children do. Transgender youth and their siblings are less quick to assume that a person's gender will remain the same over the course of their lives. Are these young people operating with a social deficit, or do they simply understand more about the fluid nature of gender? With less time on earth burdened by suffocating gender stereotypes, these youngsters may present a more holistic understanding of gender, particularly as they live in an age where gender may not serve the same purpose it once did.

Today, a small percentage of parents are raising their children in gender-neutral settings, carefully limiting positive and negative reinforcement of gender-based preferences and activities. Some of these children are being brought up without an assumed gender identity. Keeping the size and shape of genitals private, parents are using gender-neutral pronouns like they/them until a child is able to self-determine their gender as being boy, girl, or something else. Looking to this next generation may shed light on current gender identity construction and development, particularly when children are raised without the assumption that sex and gender will or should match. When these two structures are unlinked, "gender congruent" and "gender incongruent" qualifiers cease to exist. We may very soon need to look deeper into gender development models, as the systems used to measure progress and alignment may be rendered outdated.

What is gender dysphoria?

The term *gender dysphoria* is used by both mental health professionals and by transgender community members, although often in different ways. Gender dysphoria is listed in the fifth edition of the *Diagnostic and Statistical Manual of Mental Disorders* (DSM-5), published by the American Psychiatric Association. Separate from the DSM-5 diagnosis, the phrase *gender dysphoria* is sometimes used by trans community members colloquially to refer to emotional difficulty related to living in bodies or societal roles that do not fit.

The DSM-5 diagnosis of gender dysphoria is characterized by significant distress or difficulty functioning related to an incongruence between assigned sex and experienced gender identity. There is significant controversy over this DSM-5 diagnosis. Some trans people and providers advocate for continued inclusion of a gender-related diagnosis to ensure insurance payment for medical and surgical transition. However, others feel that insurance coverage can be negotiated in a different manner and that removing gender-related diagnoses from the DSM-5 is an important step in normalizing transgender identity.

How have trans people historically been approached by the medical field?

Transgender and gender-diverse populations have long been pathologized in psychiatric settings, despite the fact that a diverse array of gender identities have existed throughout cultures and over millennia. Early interventions for those who presented with trans identities typically focused on realigning gender identity with sex assigned at birth, assuming that the client was delusional or did not understand themselves as well as their provider did. Even those physicians who supported trans people typically saw them as psychiatrically ill and in need of treatment. Up until the mid-20th century, treatment included psychotherapy, shock therapy, religious training, hypnosis, lobotomy, and commitment to an asylum. Gender-affirming surgeries in the United States were illegal, and public opinion of transgender people was low. Despite these limitations, psychiatric care for trans people began to slowly change in the early 1900s with pioneering work by a few physicians.

The word *transvestite* originated in the 1910s and was coined by German sexologist Magnus Hirschfeld. Hirschfeld would later develop the Berlin Institute (also known as the Institute for Sexual Science), where the first gender-affirming surgery took place. Hirschfeld defined transvestism as the desire to express one's gender in opposition to their assigned sex at birth. Despite the now-antiquated language, Hirschfeld progressed the accessibility of gender-affirming care for his patients. He was working during an exciting time in medicine, when hormones were first being identified by western physicians. Hirschfeld offered hormone therapy and surgical interventions to help his patients achieve more authentic and satisfactory gender presentations. Most of his contemporaries at the time continued to try to "cure" individuals with cross-gender identifications.

Supporting a person's agency and decision-making process about altering their body, Hirschfeld worked to align body and mind. At his clinic, Felix Abraham began conducting what is now considered the first gender-affirming surgeries: a mastectomy on a trans man in 1926, a penectomy on a domestic servant named Dora in 1930, and a vaginoplasty on Lili Elbe, a Danish painter, in 1931. Unfortunately, much of the history of the institute's early work was destroyed in the Nazi book burnings in 1933.

Alfred Kinsey, who had studied in Hirschfeld's clinic, founded the Institute for Sex Research (now known as the Kinsey Institute) at Indiana University in 1947. In his 1948 study, *Sexual Behavior in the Human Male*, Kinsey criticized the use of the term *transsexual* as a synonym for homosexual because it implied that homosexuals were "neither male nor female, but persons of a mixed sex." After all, at that time, Americans believed that being transgender and gay were one and the same—gender identity and sexuality were not yet clearly defined as separate. The same year, Kinsey referred one of his patients, "Van," to endocrinologist Harry Benjamin, who would later standardize treatment for transgender individuals. Van, then 23 years old, was assigned male at birth and had been living as a girl since the age of three.

In 1952, the Associated Press published an article titled, "Bronx 'Boy' Is Now a Girl," announcing the medical transition of American World War II veteran Christine Jorgensen. Jorgensen's surgeon, Danish doctor Christian Hamburger, received hundreds of personal letters in the months following Jorgensen's publicity. It became clear that gender transition was not exceptional, but rather there was a significant portion of the population who felt their gender was misaligned with their bodies. Like Kinsey, Hamburger referred these clients to Benjamin.

Harry Benjamin, a German endocrinologist who emigrated to the United States, believed that those who felt their sex to be discordant from their gender deserved humane treatment in the form of hormonal therapy and affirming surgeries. By 1964, the concept of "gender identity" emerged, conceptually providing clarification that one's own understanding of their gender could be different from the sex they were assigned at birth, largely decided by visual inspection of the genitals. In 1966, Benjamin published *The Transsexual Phenomenon*, proposing a radical change in treatment: Genital reconstructive surgery, rather than religious indoctrination or lobotomization, was an appropriate treatment for those who felt their gender identity and sex assigned at birth to be at odds.

Benjamin's approach, along with his colleagues, was not universally accepted. Transgender people continued to be seen as deviant. The *Journal of the American Medical Association* published a 1978 article stating that "most gender clinics report that many applicants for surgery are actually sociopaths seeking notoriety, masochistic homosexuals, or borderline psychotics." Still, proponents of affirming care were making headway. Gender clinics, providing hormonal and surgical care (albeit only to those who met strict, heteronormative criteria) were popping up at university medical centers all over the country. In 1974, Norman Fisk popularized the concept of gender dysphoria, and in 1979, the Harry Benjamin International Gender Dysphoria Association (later renamed the World Professional Association for Transgender Health) published its first Standards of Care.

However, the same year, a study out of Johns Hopkins permanently altered the course of medical care for transgender people. At the time, Johns Hopkins had one of the leading gender clinics in the United States. Paul McHugh, then chair of the Department of Psychiatry, made it a mission to put an end to this. With his support, the clinic published a study calling sex reassignment surgeries into question by suggesting that psychosocial outcomes in transgender patients who underwent reassignment surgery were no better than those who went without surgery. Despite criticism of its methodology, McHugh and others were able to use the study to justify the closure of the clinic. Over the years that followed, the United States would see the eventual closing of nearly all university medical center gender clinics.

What sprang up in their place, perhaps slowly, but more organically, were community-based health clinics created to serve LGBTQ people. Without the backing of major universities, these early clinics were less able to offer certain aspects of transition-related care, such as surgical procedures, but, over time, private surgeons began to fill in the gaps, and, eventually, some university hospitals re-entered the arena of transgender medical care.

Although the *Diagnostic and Statistical Manual of Mental Disorders* (DSM) had not included gender-related diagnoses in its first iteration in 1952, by its second version, in 1968, a diagnosis of *transvestitism* appeared in a section called Sexual Deviations. Just as gay activists finally succeeded in removing homosexuality from the DSM, the third edition (DSM-III), in 1980, added the diagnoses transsexualism (for adults) and gender identity disorder (for children). The fourth edition (DSM-IV) followed suit, naming most gender-related diagnoses some version of gender identity disorder. For 30 years, from the 1980s to the 2010s, transgender people, whether they were psychologically distressed or not, could be given this diagnosis. It was not until 2013, with the publication of the DSM-5, that gender identity disorder was replaced by gender dysphoria, a less pathologizing name, but also a diagnosis that intentionally did not apply to all transgender people—only to those suffering from significant distress or impairment in function related to their gender identity. While, for some, this was an improvement, many transgender people continue to look forward to the day that gender-related diagnoses are completely removed from the DSM.

Even with this shift toward a more affirmative model of psychiatric care for trans people, there are still many within the field who continue to believe that transgender identity is a mental illness. Conversion therapy, also known reparative therapy, an ineffective and damaging practice that attempts to change a person's sexual orientation or gender identity, remains widespread in the United States. Although some states have moved to ban conversion therapy in minors, the majority still allow it. The Williams Institute estimates that 20,000 U.S. youth currently aged 13 to 17 will undergo conversion therapy before age 18.

Unfortunately, young psychiatrists may continue to be poorly equipped to understand transgender identity. Research shows that, on average, medical students are exposed to only five hours of LGBTQ-related content over their entire four years. During psychiatry residency, there is also no standard curriculum on these issues, leading to a workforce that is tasked with treating a population it knows little about.

What are the current controversies about transgender identity within psychiatry?

One major controversy related to transgender identity within the field of psychiatry is the role of a gender-related mental health diagnosis. The most current version of the DSM, the DSM-5, includes the diagnosis of gender dysphoria.

While less criticized than its predecessor, gender dysphoria remains a contested diagnosis. Its presence alone in a manual designed to identify and treat problematic behaviors and symptoms signals to some people an underlying cultural disapproval of transgender individuals. It is difficult for many to separate the updated gender dysphoria diagnosis from decades of pejorative and devaluing language employed by top psychiatrists and other physicians.

On a world scale, historically, the *International Classification of Diseases* (ICD) has followed a somewhat similar trajectory to the DSM, but has been quicker to depathologize transgender identities in more recent years. The eleventh edition of the ICD, published in 2018, removed gender-related diagnoses from the section on Mental and Behavioral Disorders and created a new label called *gender incongruence*, which avoids the more stigmatizing diagnosis of gender identity disorder.

The iterations of these gender diagnoses evidence many conflicting opinions from professionals in the field and community members alike. The ICD is approved by the World Health Organization (WHO). WHO is an agency of the United Nations, grounded in human rights protections and global health. As there is significant evidence that the continued pathologization of transgender identities is harmful, WHO not only has to consider objective symptom presentation but also the system by which we pathologize and the harm done in doing so.

For transgender people, a diagnosis of gender identity disorder or gender dysphoria can be a double-edged sword. Continued use of these identifiers has, for some, perpetuated harm and isolation, either self-imposed or experienced explicitly from their environment. Emotional harm aside, the diagnosis has in some cases been used as a pre-existing condition, disqualifying trans people from health care or raising their premium costs. Once identified as transgender to their insurer, many transgender people find themselves under increased scrutiny. Some trans people have reported a sudden discontinuation of insurance coverage for routine services after the diagnosis has been listed on their account. Particularly as health care systems change, there is significant anxiety with regards to what the future implications of carrying these diagnoses may be.

While times are changing, for years these diagnoses have been used without regard for clinical relevancy. Entering a doctor's office for a broken bone or sore throat, a transgender person might be diagnosed with a psychiatric disorder while undergoing no psychiatric assessment. In practice, these diagnoses have been used carelessly to identify transgender patients and not their symptom presentations. The results have been stigmatizing at best. At worst, these diagnoses can cause major problems down the line and can "out" a person as transgender without their consent.

At the same time, gender-related diagnoses are often used to justify care. For a transgender individual seeking puberty blockers or gender-affirming hormone therapy or surgeries, treating physicians typically require a diagnosis of gender dysphoria. Not only do current clinical treatment recommendations use gender dysphoria as a basis for ongoing care, most insurance plans will not cover gender-affirming treatment without documentation of gender dysphoria. Within our current healthcare systems, it is nearly impossible to provide treatment without specific documentation of a problem. When the issue is related to gender, we are presented with a conundrum: pathologize the identity or refuse care.

Not all transgender people see the diagnosis of gender dysphoria as problematic. Some people feel their incongruent bodily sex and gender identity is an issue and needs to be fixed—akin to a so-called birth defect. For others, having diagnostic criteria to draw from feels validating and affirming. For some transgender people, being able to see in written form an explanation for their emotional pain provides relief. It becomes tangible and recognizable, and there exists a course of interventions to help treat distress.

In addition to diagnostic controversies, mental health providers, especially those who routinely work in transgender care, also wrestle with their role as "gatekeepers" to medical and surgical care. While hormonal treatment is now more widely available on an informed consent basis, surgeons (and the insurance companies that fund surgeries) typically require either one or two letters of support, often from therapists or psychiatrists, to proceed. This puts mental health practitioners in the position of judging who is appropriate or not appropriate for transition-related surgeries. While a cisgender woman can arrange for her own breast augmentation without a psychiatric evaluation, the same is not true of a transgender woman.

Why do transgender people have higher rates of mental health concerns?

Without taking into account the cultural context around transgender identities, it might be easy to assume that high rates of mental health and substance abuse issues are reflective of a psychological issue inherent to this population. According to the American Psychiatric Association, children diagnosed with gender dysphoria are at higher risk of emotional and behavioral problems, including anxiety and depression. Transgender adults have increased rates of depression, substance abuse, and suicidality. Taking a step back, it becomes clear that these issues stem from years of societal stigma and discrimination, rather than from an inherent difference between transgender and cisgender people in terms of predisposition for mental illness.

Rejection that transgender people encounter is significantly harsher than the negative attitudes experienced by lesbian, gay, and bisexual youth and adults. Numerous reports point out that marginalization of transgender people from society has devastating effects on their physical and mental health. Transgender people, particularly transgender women of color, are targets of violence and abuse at higher rates than others. The subtle build of microaggressions across a lifetime, combined with outright discrimination, hostility, threats, and actual perpetrated violence, can leave a transgender person experiencing devastating symptoms associated with trauma. As such, transgender people systemically report higher levels of anxiety, depression, substance abuse, domestic violence, and homelessness. Transgender people are vulnerable to both homicide and suicide, as well as family rejection and joblessness. These negative health outcomes however, are not associated with their identity per se, but rather the profound experience of marginalization and discrimination.

Perhaps the most relevant model to contextualize transgender health disparities is minority stress theory. Minority stress theory posits that social stressors stemming from stigmatized identities account for poorer psychological functioning and compromised well-being. Health disparities among transgender individuals, in this model, can be explained in large part by stressors induced by a hostile, transphobic culture, experienced over a lifetime of harassment, maltreatment, discrimination, and victimization.

Building on earlier versions of the minority stress model, Ilan Meyer proposed a model of minority stress with lesbian, gay, and bisexual people wherein both distal stressors and proximal stressors helped to predict disproportionately high rates of psychological distress. Distal stressors were considered external to the person, such as experiences of heterosexist discrimination, and proximal stressors were internal to the person, such as awareness of stigma and internalized heterosexism. Aaron Breslow expanded Meyer's model specifically to capture the experience of transgender people. Breslow identified anti-transgender discrimination as a distal stressor, related to symptoms of psychological distress, including suicidal ideation, anxiety, and depression, in addition to poor physical health outcomes. Internalized transphobia, or the ways in which a transgender person might incorporate society's negative evaluations of transgender people into their self-concept, was a proximal stressor, which could lead to negative self-appraisals. An additional proximal stressor identified in their work was the expectation or fear of encountering future discrimination, which they called stigma awareness.

Breslow found that higher levels of minority stress (antitransgender discrimination, internalized transphobia, and stigma awareness) were associated with greater psychological distress. Resilience was found to be strongly associated with lower levels of psychological distress, suggesting that resilience may protect marginalized groups from the impact of minority stress.

There are likely many contributing factors to each trans person's individual experiences of trauma and stress. If there were questions about a child's gender identity when they were young, a parent's ability to manage their own reactions could have impacted the young person's development, their sense of safety, self-acceptance, and ability to think constructively about their future. Rejection of identity by a caregiver or forced identification as cisgender can set up a child for feelings of guilt and worthlessness. They may grow up incorporating feelings about themselves as being wrong, unnatural, or sinful. If confronted with bullying at school, their ability to connect with peers and socialize can be hindered. They could be preoccupied in classes, distracted by anticipation of being misgendered by their classmates or teachers, or may feel unwelcome or unsafe in their school, all factors that contribute to their overall academic achievement.

Because societal rejection, rather than something innate to transgender people, is what leads to poor mental health outcomes, there is hope that future generations will have improved mental health. In fact, recent research published in the journal *Pediatrics* shows that trans children who grow up in supportive environments have similar levels of depression and anxiety as their cisgender peers. Numerous studies of trans adults demonstrate that quality of life improves and depression and anxiety decrease with access to affirming medical and surgical care.

What do medical and surgical options look like for transgender people?

Medical and surgical interventions play an important role in the transition process for many transgender people. However, there are social changes that can also help trans people to feel more comfortable in their bodies and social roles. For some, "social transition" is an extremely therapeutic step. For some trans people, social changes are the focus of transition, as they may not desire medical or surgical treatment. Social transition can include changing pronouns, name, haircut, and style of dress. It may involve updating documentation to reflect the person's gender identity. These social changes can help integrate the person into their world in a more desirable and authentic way and decrease the amount of time they spend internally reacting or adjusting to being called a name or by a pronoun that feels in their core inappropriate, or even in opposition, to their gender.

In addition to social changes, many trans people also desire some form of medical and/or surgical treatment. If assigned female at birth and currently identifying as male, masculinizing interventions such as taking testosterone can help someone feel more at ease in their body, as well as operate and function in the world as male. The same is true of feminizing hormones (typically estrogen and spironolactone in the United States) for those assigned male and identifying as female. Nonbinary individuals may choose to take hormones as well, sometimes, but not always, in lower doses or for a defined period of time. Masculinizing hormone therapies can change hair growth patterns, vocal range, muscle and fat distribution, along with suppressing menstruation and changing the appearance of the genitals. Feminizing hormone therapies are used to elicit breast growth and alter fat distribution. They can also block the effects of testosterone, including halting the progression of male pattern baldness, and decreasing spontaneous erections and testicular volume.

Numerous surgical interventions are available to masculinize and feminize bodies. Breast tissue can be reduced or enlarged. Genital structures can be modified to resemble a typical penis or vagina. Surgical contours can be made to the body and face that help to signal masculinity or femininity. Both behavioral and surgical interventions can assist with developing a more masculine or feminine voice.

When gender identity and body appearance are aligned, there is usually less distress. However, not all bodies can be medically and surgically altered to reflect an internal gender identity, and not all people desire to seamlessly "blend in" with cisnormative ideals. For nonbinary individuals, matching their outside physical body to their internal identity can mean incorporating both masculine and feminine markers of gender. Their goal may not be simply to "pass" as a cisgender man or woman, but rather to express authentic qualities of both. For any individual transgender person, there may be a variety of reasons they choose not to engage in an "all or nothing" approach to medical transition. They may simply not identify with or want certain changes, find costs prohibitive, or not be able to undergo such invasive procedures due to other unrelated medical concerns.

Can transgender people have biological children?

In 2008, the sensational story of Thomas Beatie, "the pregnant man," burst onto news networks across the country. Beatie, a transgender man, made appearances on *Oprah*, *Larry King*, *The View*, and *Good Morning America*. People magazine captured his life in a six-page photo spread. Beatie explained that his wife was unable to become pregnant, and so he chose to do so instead.

For many Americans, Beatie was the first man they had ever heard of who was pregnant or had given birth. However, many transgender men have children every year. The numbers are difficult to estimate, but their stories have spawned multiple news articles and documentaries.

Surprising to most people is the fact that testosterone does not generally cause permanent infertility in trans men. While continued fertility is not a guarantee, many trans men have started testosterone and then stopped it later, resumed menstruation, and become pregnant. Some trans men have even had accidental pregnancies while on testosterone if the dose was not high enough to prevent ovulation. Because testosterone can cause birth defects, it is recommended that all trans men on testosterone who are at risk of becoming pregnant use a form of birth control, and that trans men trying to conceive stop taking testosterone.

Many surgeries that trans men undergo do not remove the possibility of later giving birth. The most common type of surgery amongst trans men is "top surgery" (the creation of a male-contoured chest). This type of surgery does not interfere with reproductive capacity. It does decrease breast tissue and often removes milk ducts; however, there are many different variations of "top surgery," and some trans men are able to "chestfeed," even after top surgery.

Besides becoming pregnant themselves, there are other ways for trans men to have children that are biologically connected to them. Many trans men have harvested eggs and used in vitro fertilization to implant embryos in a partner who later gave birth. Some trans men decide to preserve eggs or embryos for later use. Either of these methods tends to be more expensive than becoming pregnant via natural conception or sperm donation. However, they may allow for the possibility of the trans man not carrying the pregnancy to term himself.

For some trans men, the idea of becoming pregnant and giving birth is uncomfortable. Pregnancy is considered by most people to be a feminine activity. In a 2014 survey of trans men who had become pregnant, many of the men purposefully chose words like "dad," "carrier," and "gestational parent" to describe themselves in masculine or gender-neutral ways. Some of the men discussed their reasons for becoming pregnant in exclusively pragmatic terms, saying things like, "I looked at it as something to endure to have a child" and "My body was a workshop, building up this little kid."

Becoming pregnant as a trans man can bring up feelings of intense gender dysphoria, and many trans men avoid it for this reason. However, others don't have as many contradictory feelings. It can be difficult for pregnant trans men to interact in social situations, as many of the people they encounter will have little experience with male pregnancy and can even become hostile or aggressive. In the end, the choice about whether to become pregnant is a personal one but influenced by the society in which we live.

So far, no transgender women have yet been able to become pregnant. Scientifically, many predict that this will one day be possible. In fact, a handful of cisgender women born without a uterus have now given birth after uterine transplants, the first in 2014.

Transgender women who would like to have biological children generally do so either through natural conception prior to beginning hormone treatment or by preserving sperm, which is later used by a partner or a surrogate. Unlike trans men, trans women do not generally have the option of starting hormone therapy and then choosing fertility options, as feminizing hormones can lead to rapid irreversible infertility. For this reason, trans women who are at all considering having biological children are often encouraged to think about sperm-banking prior to initiating hormones.

Like trans men, trans women may have the possibility of breastfeeding, although this practice is fairly new, at least within the medical field. The first case study of a health center inducing lactation in a transgender woman was published in 2018.

It is likely that as the field of transgender health progresses, more options will become available for transgender people to pursue various ways of creating families.

What role do mental health professionals play in the field of transgender health?

Many transgender people cope well with the stress of living with societally marginalized identities. However, there is much to be gained from working with mental health practitioners who have experience helping to navigate gender-related stress.

Even when fully cleared for medical transition, many people who seek intervention to alter their physical body with the hopes of alleviating gender dysphoria may benefit from mental health support to help explore the process. Like any major medical change, processing risk, managing anxiety, and preparing aftercare are all vital when undergoing such a major step, in addition to thinking critically about what aspects of life are likely to change with medical transition and what will remain the same despite such alterations. Mental health providers who treat trans clients therefore often explore with them the psychological implications of gender related interventions, as well as help them concretely plan to ensure ideal outcomes.

A therapist may also help someone manage living with gender dysphoria, particularly when medical and surgical interventions do not fully resolve the difficulties of living in a body or societal role that feels inappropriate to a person's identity. These clinicians may help people increase their tolerance of distress, provide space to explore the evolution of one's gender identity, and work toward creating opportunities for acceptance, support, and validation.

Transgender people can benefit immensely from community support, particularly when early in the process of transition. Especially helpful are support groups, community conferences, and educational events wherein information and resource sharing is the primary focus. Other social interventions may include participating in transgender cultural events, socializing on a sports team or in a theater group, linking up with a mentor who has been through a similar process, or even participating in an online community or forum. The goal of these interventions is primarily to decrease isolation.

There is no one-size-fits-all approach to treating gender dysphoria, due to the variety of ways in which people experience distress around their gender, as well as the diversity of gender, gender identity, and gender expression. Youth may benefit from support around social and legal transition to change their name and legal gender to reflect a lived gender identity rather than what was assigned at birth. They may also respond well to affirming individual psychotherapy, family therapy, support around gender transitions at school, puberty-suppressing hormones (to block the masculinizing and feminizing byproducts of puberty and reduce the need for surgical interventions later on) and cross hormone therapies, and possible surgical interventions when mature. Adults who have passed their own puberty are generally not appropriate candidates for puberty suppression medication, but benefit from many of the same interventions, along with couples counseling and support around workplace transitions.

Current best practice for mental health practitioners in the treatment of gender dyphoria is to take an affirmative approach that views trans identity not as a pathology but a normal human variation. As such, treatment focuses on reducing the anxiety and depression associated with pervasive transphobia and an experience of feeling "othered" in social settings. Affirming approaches generally suggest mental health treatment be offered but not enforced or mandated; highlight education for caregivers, friends, and family; and encourage community engagement.

Affirming interventions come in stark contrast to conversion therapies, which work to undo any cross-gender identifications. These techniques have been shown to be both harmful and unsuccessful. Gender identity does not appear to be malleable, and therapy aimed to change it typically leads to increased trauma and overall distress. Current research supports inclusive therapies, finding that affirmative approaches can decrease depression and suicidality. These findings are supported by most ethics boards, which generally condemn conversion therapies and ask clinicians to find ways to support their transgender clients, be it via traditional talk therapy modalities or by linking them to medical providers who can help them access gender-affirming interventions.

Why is access to care an issue for transgender people?

Transgender people often benefit from medical and mental health interventions to alter their bodies to reflect more authentically their lived gender identities and to process what can be an overwhelming experience of transphobia. Yet many find it nearly impossible to get the care they need to feel better.

The United States Transgender Survey, most recently conducted in 2015, found that 86% of respondents reported they were covered by a health insurance or health coverage plan. The remaining 14% were uninsured, which was slightly more than the 11% of the general U.S. population who were uninsured the same year. Of note, however, is the discrepancy existing between transgender and cisgender individuals' utilization of healthcare. While less likely than the average American to be insured, transgender people require access to medical systems for aspects of their daily lives in ways that others do not. If taking hormones, they may need daily medication and routine blood work to ensure optimal functioning. Preventive care is particularly important as this current generation of transgender people age, particularly given the lack of research and longitudinal data available for long term use of hormone therapies.

In the United States Transgender Survey, insurance coverage differed by region, race, and immigration status. Those in the South were uninsured at a rate of 20%, compared with 13% in the Midwest, 11% in the West, and 9% in the Northeast. Among people of color, Black transgender people in the United States were uninsured at a rate of 20%, compared with 18% of American Indian respondents, and 17% of Latinx individuals. Asian and Middle Eastern respondents were uninsured at a rate of 11%, and white respondents, at a rate of 12%. Those who were not U.S. citizens were more likely to be uninsured, including nearly one fourth (24%) of documented noncitizens and 58% of undocumented residents.

For those individuals who were insured, over the past year, one in four respondents experienced a problem with their insurance related to being transgender, such as being denied coverage for care related to gender transition or being denied coverage for routine care because they were transgender. More than half of respondents who sought coverage for transition-related surgery in the past year were denied, and one fourth of those who sought coverage for hormones in the past year were denied.

One third of transgender respondents who saw a healthcare provider in the past year reported having at least one negative experience related to being transgender, with higher rates among people of color and people with disabilities. Issues included being refused treatment, verbally harassed, or physically or sexually assaulted. Many had to teach their providers about transgender people to get appropriate care. Unsurprisingly, many respondents did not see a doctor when they needed to—23% because of fear of being mistreated and 33% because they could not afford it.

These concerning statistics reveal systemic inequities. The best insurance policies are offered to those who have steady employment, usually necessitating one be educated at a certain level, an impossibility for some young transgender people facing unbearable discrimination in their homes and schools. Among transgender people, family rejection is associated with increased odds of suicide attempt and substance abuse. To think systematically about increasing one's chances for optimal health insurance coverage, we must consider early childhood environment and every subsequent experience as cumulatively affecting chances for inclusion, success, and ability to access appropriate care.

Even for those trans people who are able to see medical and mental health providers, experiences may vary significantly. Education about trans care is routinely left out of graduate training programs, and clinicians typically lack training and core competencies needed to appropriately care for a transgender person. This lack of education renders even the most well-intentioned clinicians at risk of saying something harmful to their transgender client. Negative experiences in healthcare settings may reduce a transgender person's actual or perceived ability to get their current needs met. Access to care is confounded by numerous factors not limited to financial inequalities; lack of access to insurance coverage and specialists in the field; and lifelong experiences of transphobia, rejection, and pathologization in the home, school, and social spheres.

Intersectional experiences that force someone to navigate systemic racism, ableism, and xenophobia can further complicate their lifelong success in accessing affirming and inclusive healthcare.

What role has gender played in the diagnosis and treatment of mental health concerns?

Colloquially, the word *hysteria* is used to describe excess emotion, out of proportion to life events. Historically, when used by doctors, it has referred to women who report physical symptoms without a known medical cause, although it was often broadly applied to women who presented with anxiety, depression, and other mental health concerns. Hippocrates, a Greek physician in the fifth century BCE, was the first to use the term *hysteria*, believing that these types of ailments were linked to the movement and placement of the uterus, or *hystera*, in Greek. Well into the 19th century, hysteria continued to be thought of as a result of problems with women's sexual or reproductive functioning; treatments included marriage and orgasms.

In the late 19th century, symptoms of hysteria, or what we today call conversion or somatization, began to be linked to trauma. One of the leading proponents of this theory was French neurologist Jean-Martin Charcot, who argued that men could also have somatic symptoms, particularly if they had experienced traumatic events. Suddenly, a disorder that had been thought of, for millennia, as being the fault of women's inferior bodies, was now potentially linked to a force outside of them.

Today, women are 75% more likely than men to be diagnosed with borderline personality disorder (BPD), a chronic condition characterized by mood instability, feelings of emptiness, fear of abandonment, suicidality, and self-harming behaviors. BPD is thought to be directly linked to trauma, particularly what is called "complex" trauma, which involves multiple traumatic events, often early in life and usually interpersonal, such as sexual abuse, which is more common in girls and women. Despite its roots in trauma, BPD is a stigmatized diagnosis. Those with this condition are thought of as difficult, irritational, and manipulative. Even many clinicians take the stance that clients with BPD are responsible for their symptoms as part of a moral failing, rather than understanding that this condition is a result of trauma. In her book *Women and Borderline Personality Disorder: Symptoms and Stories*, author Janet Wirth-Cauchon writes that "the label 'borderline' may function in the same way that 'hysteria' did in the late 19th and early 20th century as a label for women" (p. 8).

The history of hysteria (and BPD today) is emblematic of the way western societies, including their physicians, have approached women's emotional lives. Even today, women's medical and mental health concerns are often brushed aside, and women are painted as overreacting and overly emotional in comparison to men, who are thought of as more motivated by reason and more in control of themselves. Women's mental health issues also continue to be blamed on something inside of them, rather than looking for societal causes.

5

GENDER, SOCIETY, AND BEHAVIOR

What is the social construction of gender?

Imagine if you were a visitor from another planet and someone handed you a $100 bill. Would you recognize it as a gift? As money? How would you know if it was worth a large or a small amount?

The meaning we assign to objects, people, and events as a society is also known as social constructionism. By comparing different societies, we can see that what we think of as objective reality is often socially constructed. A visitor from another planet is unlikely to understand our money until we explain it. What makes money valuable is our collective belief that it is.

Most scholars believe that some portion of our understanding of gender is socially constructed. That is, as a society we have created gender roles that we agree on and perceive as "real," despite them not being based on objective differences between women and men.

There is significant evidence supporting the idea that gender is at least partially socially constructed. Comparing men and women across the globe today, we can see that what is considered masculine or feminine is not universally accepted. Men, for example, in some areas of the world wear what would be considered dresses and therefore unacceptable in other areas. Men kissing and holding hands with each other is part of expected interaction in some countries, while in others it is viewed as feminine. Even in the United States, over time we have seen drastic changes in gendered expectations, with women entering the workforce and men spending more time parenting.

In the 1960s, as feminist movements started to take shape, researchers began to explore the ways in which male and female children were raised. They made what would today be considered a not very surprising discovery—that adults treat children differently based on their gender. Adults in their studies used more sex-stereotyped toys with girls and bought boys more toys that elicit competence. They treated girls more gently and punished boys more physically. They allowed boys more freedom to roam. Adults also perceived children's emotions differently based on their gender. In one study, participants were shown a video of a baby playing with a jack-in-the-box. Those who were told that the baby was male were more likely to describe the baby's reactions as "angry" and those who were told that the baby was female were more likely to describe the same reactions as "fearful."

While most people would agree that society has at least some influence on gender roles, there are those who go even further, arguing that gender is a completely social construct—that male and female bodies are different, but the meaning assigned to them stems from society. One of the best known of the social constructionists is Judith Butler, a feminist philosopher who opposes the idea of "core" traits of particular genders. Instead, she proposes that, as a society, we continuously recreate gender roles by performing them and that the categories "men" and "women" would not exist without this performance. A common misunderstanding of Butler is that gender performance is similar to acting in a play and that we can therefore easily throw off the oppressive chains of our assigned genders by simply dressing or behaving differently. Instead, Butler is arguing that our repetition of gendered performances is part of our societal construction of gender and a loop we are all stuck in, in some way or another.

Do some cultures have more than two gender options?

In many countries today there is growing visibility and conversation around nonbinary gender identities. In 2016 Oregon resident Jamie Shupe became the first U.S. citizen to be legally recognized as nonbinary, with many following in Jamie's footsteps and legally updating their legal gender markers from "M" or "F" to "X" or "Unknown."

Some individuals have gender identities that lie between the boundaries of "man" and "woman." Many of these individuals would be categorized under the umbrella term *transgender* and may use language like *nonbinary, genderfluid, agender, genderqueer, bigender,* or *gender nonconforming*. In 2014, Facebook began allowing users to choose among 58 defined genders, with many other social media networks and online dating sites following their lead. Today, many university applications have options for gender identities that fall beyond male and female. There are third gender or "other" gender options for some local government records, as well as for many organizations and programs providing social services and aid. However, while there is increasing accessibility for an individual to change their legal gender from male to female or female to male, at the time of this writing it is still difficult if not impossible in most U.S. states to legally be recognized as a third gender.

Despite legal hurdles, there is an undeniable growing conversation and acceptance of gender diversity. Schools and social service agencies are urged to be mindful of inclusion of nonbinary identities, based in large part in evidence of the psychological and educational roadblocks associated with disregard of such vital aspects of personhood. But there are those too who stand in fierce opposition to policy change that would increase acknowledgment of nonbinary and intersex individuals (those who are born with bodies not considered either completely male or female). Objectors may have religious or political doubts regarding the legitimacy and realness of such identities. Pathologized, nonbinary and intersex bodies are often seen as defective, and gender identities invisible to the human eye continue to be regarded with skepticism and doubt. Others have voiced concern that seeking legal gender reassignment could make someone more difficult to trace or track, and could increase the potential for identity fraud or evading a warrant or search.

While it is true that creating a legal record of transgender and nonbinary people who fall outside the traditional binary necessitates a major overhaul in systems (and social awareness), volumes of evidence indicate that individuals with genders apart from "male" and "female" have existed throughout time and across the globe. Some of our oldest records, gleamed from ancient Egyptian pottery created between 2000 and 1800 BCE, depict three distinct genders: tai (male), sḫt (sekhet), and hmt (female). In Mesopotamian mythology, among the earliest written records of humanity, there are references to types of people who are neither male nor female. Plato's *Symposium*, written around the 4th century BCE, details a creation myth involving three original sexes: female, male, and androgynous. Rabbi Elliot Kukla, who studies religious scripture and gender, has commented on the existence of six distinct genders referenced in ancient religious texts. Kulka finds the word *androgynos*, referring to a person with both male and female sex characteristics, used 149 times in the Mishna and Talmud (1st–8th centuries CE) and 350 times in classical midrash and Jewish law codes (2nd–16th centuries CE).

The term *Two Spirit* is a phrase that has been adopted to identify a range of nonbinary identities existing across over 150 indigenous tribes of North America. Two Spirit individuals were those thought to possess the qualities of the sun and the moon, man and woman, and were sometimes elevated in social class and function. Some considered them gifted in their ability to understand multiple elements of community life and culture. Algonquian, Navajo, Lakota, and Ojibwe languages have multiple words used to identify individuals with distinct gender identities. This includes the Navajo identity of nádleeh, referring to an "effeminate male" or someone who is "half woman, half man," and the Lakota wíŋkte, who hold a separate social category in Lakota culture. Wíŋkte are generally thought of as male-bodied people who adopt the clothing, work, and mannerisms that Lakota culture usually consider feminine. While historical accounts of their status vary widely, most accounts see the wíŋkte as full members of the community, not marginalized or ostracized. Other accounts hold the wíŋkte as sacred, occupying a liminal, third gender role in the culture and born to fulfill ceremonial roles that cannot be filled by either men or women.

In the Zapotec communities of Oaxaca, Mexico, muxes are a third gender of individuals who exist outside of the gender binary. Muxes are typically assigned male at birth and identify with a more feminine gender identity or exist wholly outside the male–female binary. One 1974 study estimated that 6% of those assigned male at birth in an Isthmus Zapotec community were muxe. Muxes hold different social roles than the men and women in their communities, serving distinct and vital positions in family life, elder care, and within the arts. Anthropologists believe the acceptance of people of mixed gender can be traced to pre-Columbian Mexico, where accounts exist of Aztec priests and Mayan gods who wore mixed gender garments and were considered and revered as both male and female. Since the 1970s, Oaxaca holds an annual festival, Vela de las Intrepidas, in their honor.

In Native Hawaiian communities, Māhū describes an individual who embodies both male and female spirits equally. In the precolonial history of Hawaii, Māhū were priests and healers who passed down genealogies and cultural traditions, performed sacred hula dances and chants, and were positioned as the person in the community parents would ask to name their children.

Fa'afafine is the word used to describe a third gender category amongst the American Samoans. A recognized third gender, fa'afafine are assigned male at birth and explicitly embody both masculine and feminine gender traits. Of note, Polynesian culture is altogether less binary than western standards as it pertains to gender and sexuality.

In South Sulawesi, Indonesia, the Bugis people recognize five different genders: makkunrai, oroané, bissu, calabai, and calalai. Makkunrai and oroané are comparable to cisgender men and women. Bissu, calabai, and calalai are nonbinary genders that are less comparable to western standards: Bissu are mixed gender, or "meta-gender" priests and shamens, while calabai and calalai are similar to trans women and trans men, respectively.

In Southeast Asia, hijras are individuals not considered either male or female. The word *hijra* is a Hindi/Urdu word, derived from the Semitic Arabic root *hjr,* meaning "leaving one's tribe." With a recorded history of over 4,000 years, hijras have a complex history of being both celebrated and criminalized, and were especially denigrated during British rule. In 2014 India began legally recognizing hijra as a third gender, following similar legislation in Nepal, Pakistan, and Bangladesh.

Thai kathoey live similarly to the western concept of transgender women, although not all transgender women in Thailand identify as kathoey. The identity is tied to spirituality and religion, as kathoey are viewed as conduits for both male and female spirits. In ancient Thai Buddhism, four genders were acknowledged: male, female, bhatobyanjuanaka, and pandaka.

Third genders also exist in Pakistan, where Khawaja Sira are both woven into spiritual customs as revered leaders and simultaneously harshly discriminated against. In Oman, Xanith identities largely refer to individuals who were male assigned at birth but who live as women. The ergi of Siberia;

the sekrata of Madagascar; the ashtime of Ethiopia; and the binabae, bayot, agi, bantut, badíng, and the lakin-on of the Philippines are all third gender categories that are culturally acknowledged.

Today, throughout the globe, these nonbinary and third gender categories are integrated into society in often precarious roles. Many are linked to spiritual enlightenment and revered in positions of leadership, while at the same time they face violence, human rights violations, and loss of family of origin, forcing them to relocate into their own self-contained communities.

How does language shape the way we think about gender?

Do people who speak different languages see the world in different ways? According to the Sapir–Whorf hypothesis, also known as linguistic relativity, they do. Those who support the theory of linguistic relativity argue that language can shape the way we think and even limit the things we are able to think about. Charlemagne is said to have stated, "To know a second language is to possess a second soul."

One of the most commonly cited examples of linguistic relativity is the idea that Inuit people are able to understand and describe snow in more ways than others because their languages have more words for it. However, in 1991, one linguist dubbed this the "great Eskimo vocabulary hoax," arguing that there was little evidence that there were concepts related to snow that Inuits could express that others could not. The debate continues, with the 2010 publication of the book *Knowing Our Ice*, which asserts that there are, in fact, specialized vocabularies for snow in Inuit languages and that others do not have access to as rich a world in this arena.

There has been significant controversy around the concept of linguistic relativity, with linguists and anthropologists debating whether the languages we speak actually limit us cognitively. Even if virtually all languages are translatable, does our own language still shape our world view? While most experts oppose the suggestion of linguistic determinism (the idea that the language we speak entirely determines our possible thoughts), many feel that there is evidence for linguistic influence (the idea that our language can provide constraints on our thinking).

There are some languages, such as Balinese, for example, where instead of using terms like *left* or *right*, speakers always refer to cardinal directions (i.e., *north* or *east*). There appear to be differences in the way people who speak languages like English, which uses relative directions, and those who speak languages with cardinal directions orient themselves in space. Perhaps even more interestingly, spatial orientation can also affect how we think about time. The Kukutai, for instance, a group of aboriginal Australians whose language uses cardinal directions, perceive time as moving from west to east rather than left to right, so if they are facing south, they think about time as moving from right to left rather than left to right.

Do the languages we speak have the potential to shape the way we think about gender? Research shows that they may. The most obvious example of this is the use of sexist language. Children asked to talk about or draw pictures of people who do certain jobs are more likely to suggest the person is a man if the job title is sexist (i.e., fireman, policeman, spokesman).

Some languages have "grammatical genders," meaning that nouns are masculine, feminine, or neuter. Many English speakers learning other languages have difficulties remembering the genders of nouns because English does not have grammatical genders. Why is the word *cat* (el gato) masculine and the word *table* (la mesa) feminine in Spanish? Why, in Greek, is the word for "sun" (o helios) masculine and the word for "moon" (to fengari) feminine, but the opposite is true in German (the word for "sun" is feminine [die sonne], and the word for "moon" is masculine [der mond])?

While it may not appear to be important whether a word is masculine or feminine, grammatical genders may subtly affect the way we think. Studies have shown that the adjectives people use to describe objects depend on the grammatical gender of the object in their native language. When asked to describe a key (masculine in German and feminine in Spanish), German speakers use more stereotypically masculine words such as *hard, heavy,* and *useful,* while Spanish speakers use more stereotypically feminine words such as *intricate, little,* and *shiny.*

Although only some languages employ grammatical genders for nouns, most (but not all) use gendered language to talk about people. In English, pronouns have traditionally been gendered (i.e., *he* or *she*), making it difficult to talk about someone without referring to their gender. In languages without gendered pronouns, such as Finnish and Japanese, you do not have to know the gender of the person you are speaking about to talk about them. Does this mean that Japanese speakers do not think as much about the gender of the people they are describing as we do? If so, do Japanese speakers have a different gendered world view from ours? It is hard to know.

In some languages, gendered words are necessary not just when referring to others, but also when describing yourself. A man would be expected to refer to himself as "grand" (tall) in French, for example, while a woman would be expected to refer to herself as "grande." Children who grow up speaking French learn early on to talk about themselves in gendered ways. Two children on a playground in an English-speaking community may be able to have an entire conversation without knowing each other's genders. This would be much more difficult for the same children speaking in French. When transgender people transition in French-speaking communities, they must change the genders of the adjectives they have been using to describe themselves since childhood. French is not unique; many languages including Hebrew and Arabic require a speaker to conjugate words dependent on the speaker's gender.

Many languages use gendered language to refer to groups as well as individuals, and gendered language for groups is often sexist, favoring men. In Spanish, for example, a group of women would describe themselves as the feminine *nosotras* (we), but a mixed group of women and men, no matter how few men were present, would be expected to describe themselves as *nosotros* (masculine). In English, we regularly describe mixed-gender groups as "you guys" but would be looked at strangely for calling the same people "you gals," even if there were just one man in the group. Why is "you gals" so jarring? Is it simply because it would be incorrect linguistically? Another possibility is that the convention of referring to mixed groups as masculine has contributed to a hierarchy in our minds, making us uncomfortable with the idea of referring to a man as feminine. Which explanation is correct? How can we ever truly know?

How has culture shaped language about gender?

In subtle ways, our language likely affects the way that we think. But how does the way that we think affect our language? In other words, how does our culture shape the way our language develops? And, more specifically, how does our cultural understanding of gender affect the language that we use?

There are certainly some obvious examples of ways in which culture affects gendered language. In patriarchal societies, women are often referred to using words that are rarely, if ever, used to describe men, and vice versa. Women are beautiful, while men are handsome. Women who take charge are bossy, while men are strong leaders. When was the last time you heard a man referred to as bubbly, curvy, ditsy, frigid, frumpy, or shrill?

Culture also shapes the way men and women talk. A stark example is a Native American language called Koasati, in which there are clearly demarcated differences between male and female speech. Women end certain words with vowels where men would use a consonant, and women and men use different pitches and stresses. In some ways, this is not so different from English, where men and women often have different intonations. In English, women have also been found to use more polite language and to ask more questions. When analyses are done, it is clear that this is not because women necessarily want to gain more information, but because they use questions to engage the other person. Researchers have found that men are more likely to use language for status, while women use it to connect and create intimacy.

Language is not just about the words that we use or how we say them. It also involves the amount of space that we give to others to speak and express themselves, and how well we listen (or don't) to each other. Women regularly bring up the fact that men are more likely to interrupt them, and a recent study showed that this occurs even at the highest levels of society. Through reviews of transcripts of Supreme Court oral arguments, a research group showed that, in 2015, 65.9% of all interruptions were directed at the three female justices (out of nine total justices). Despite strict rules that justices can interrupt each other, but lawyers are never, for any reason, permitted to interrupt justices, male lawyers accounted for 10% of interruptions of justices. (Female lawyers interrupted justices so few times that their interruptions statistically neared 0%). The most common form of interruption of any justice was a male lawyer interrupting the only woman of color on the court, Justice Sonia Sotomayor.

Although culture can have a negative impact on language and perpetuate stereotypes, language can also be reclaimed and purposefully shifted and changed to create new frames and ways of thinking about things, including gender. For many years, the default pronoun in English, when gender was not known, was *he*. Feminists worked to shift this default to *he or she* or sometimes to just *she*. When the idea to expand the default pronoun was first suggested, it was met with significant resistance, with many people, including some women, arguing that it was too difficult to change entrenched language. However, starting only a few decades later, writers were considered old-fashioned if they used the default *he*. Similarly, when feminists first suggested using neutral terms like *spokesperson* and *firefighter* instead of more gendered language, there were complaints that it would be hard to make this transition, but now these more neutral terms are normal parts of our vocabulary.

Today, one of the areas of the English language that appears to be changing rather rapidly is self-defined terminology that refers to groups of people, including language related to race, ethnicity, gender, sexuality, immigrant status, and disability. Many people are eager to learn and use language that is considered respectful to others. With regard to gender, in addition to the categories of male and female, there is now more widespread use of terms like *nonbinary*, *gender expansive*, and *genderqueer*. It is also becoming more common for people outside of LGBTQ communities to understand and use gender-neutral pronouns such as *they* to refer to one person.

Recently, native speakers of a number of languages have attempted to incorporate more gender-neutral ways of speaking and writing. In Swedish, traditional pronouns are *han* (he) and *hon* (she). Over the last 5 to 10 years, the gender-neutral pronoun *hen* was introduced. Likely because of a combination of interest and ease of use, *hen* has taken off. In Spanish, many people have begun to use the letter *x* in place of the masculine *o* or feminine *a*. The word *todos* (meaning "everyone") in the masculine form, for example, would become *todxs* and could be pronounced "todes." In the United States, *Latinx* is now a popular self-description for those who may have called themselves *Latino* or *Latina* in the past. The *x* in *Latinx* is typically pronounced like an English *x*.

English, like Spanish, does not have as easy of an answer to the question of gender-neutral pronouns as some other languages, and using a formerly plural pronoun as singular can initially feel somewhat clunky. However, use of the pronoun *they* to describe one person also seems to be catching on.

The only thing constant about language is that it evolves. Sometimes this takes place organically, making it easy for stereotypes and biases to enter into it. However, at other points, groups of people work hard to mold language in ways that help us to communicate more respectfully and to build a more just world.

What are gender-neutral titles and pronouns?

Most people think that the title Ms., which, like Mr., reveals nothing about the marital status of the person it describes, was coined in the 1960s during a time of feminist movements for social change. In fact, its use first began in the 17th century, but fell out of favor. During the 20th century, there were at least a few who attempted to bring Ms. back, arguing that it was more convenient to use and that it could be embarrassing to call a married woman *Miss* or a single woman, *Mrs.* However, conventions can be difficult to overcome. Finally, it was feminists in the 1960s, most notably Gloria Steinem, who named her 1972 magazine *Ms.*, that popularized the title.

Today, men continue to have only one title (Mr.), while women can choose from three (Mrs., Miss, or Ms.). Recently, some transgender and nonbinary people have begun to use gender-neutral titles, and it is possible these will come into more mainstream use by cisgender people at some point as well. Mx., pronounced "mix," is the most common gender neutral title.

In English, in addition to titles, pronouns are also gendered. We typically use the pronouns *he/him* to describe men and *she/her* to describe women. Many transgender people identify with these binary terms, but there are also those who would prefer to be addressed in more gender-neutral ways.

There were early attempts to create new words, such as *ze* to replace *she* or *he* and *hir* to replace *him* or *her*. There have also been some unique gender-neutral pronouns that have emerged, including *yo*, used by teens in Baltimore. By far the most common gender neutral pronoun at present is *they/them*. While English teachers may complain that it is grating to their ears, it is a word people know and are used to using, even if not in the same context, and has become more and more commonplace in the last decade. While *Miriam-Webster Dictionary* only recently added *they* as a valid singular gender-neutral pronoun, interestingly, the word *they* being used in this way isn't entirely new. Examples of the singular *they* being used to describe one person can be seen as early as 1386 in Geoffrey Chaucer's *The Canterbury Tales*, as well as Shakespeare's *Hamlet* in 1599. Even in modern times, without thinking twice, we might remark, "Someone left their coat on my chair," unintentionally resorting to a singular gender neutral pronoun when not knowing the gender of the person in question.

How do different major religions understand gender?

The majority of the world population identifies as religious. In the United States, surveys indicate that somewhere around 80% of participants belong to an established faith, primarily Christianity (70%), with much smaller proportions of other major world religions (Judaism 2%, Islam 1%, Hinduism 1%, Buddhism 1%). On a global level, there are an estimated 2.3 billion Christians, 1.8 billion Muslims, 1.1. billion Hindus, and 500 million Buddhists. Jews make up a much smaller percentage of the world population (15 million total) than the United States population, and there are other religious groups less well known in the United States whose numbers are higher on a global scale (e.g., 27 million Sikhs).

Women in the United States are more likely than men to say religion is "very important" in their lives (60% vs. 47%). American women are also more likely than American men to say they pray daily (64% vs. 47%) and attend religious services at least once a week (40% vs. 32%). Despite women's interest and participation in religion, many sects of major world religions separate participation in religious rituals and services by gender and exclude women from senior roles. Queer and trans people have also historically been marginalized, and often condemned, by major world religions.

To make blanket statements about the role of gender or sexuality in any one religion is nearly impossible, as there are progressive and conservative branches of all major world religions. In addition, there are innumerable dimensions by which each religion could be measured in terms of its approach to gender and sexuality. Important questions include: Who are the major historical figures, saints, and gods? Are they male, female, or beyond gender? Do the religious texts promote gender equality or condemn women, queer, and trans people? Are women, queer, and trans people permitted roles in the clergy, and can they attain the same status within the religious hierarchy? Are there current policies within the religion that are inclusive of women's and queer/trans rights?

Although the ancient Hebrew texts referred to God in some places as genderless or, alternately, male and female, in the Abrahamic religions (Christianity, Islam, Judaism), most of the major prophets and historical figures have been men, and God is often referred to as male (the Father). Female figures like Deborah, the first judge in Hebrew tradition; Fatima, Muhammad's wife; and a number of female saints in Catholicism are notable exceptions. In Hinduism, the Brahman, or Supreme Self, is a genderless concept. Hinduism is polytheistic, and its deities come in many genders, including male (e.g., Krishna), female (e.g., Lakshmi), and androgynous (e.g., Ardhanarishvara). In addition, Hijras, a third gender category in India and Pakistan, are often thought of as spiritual beings and asked for blessings. Still, the majority of well-known historical figures within Hinduism are male. The founder of Buddhism, Siddhartha Gautama, was male, as well as most of the influential teachers, although there have also been female buddhas and spiritual leaders. A number of indigenous religions in the Americas included shamans who were gender nonconforming people.

In addition to the genders of gods and historical figures, a religion's approach to gender can also be garnered from its texts and the teachings of its early leaders. Sikh scriptures are some of the most straightforward in their support of women, outlining the equality of men and women, who are described as having the same souls, with temporary outward appearances. The early Sikh Gurus also made many statements in support of women's rights, instructing followers that women were to be equal participants in religious ceremonies and condemning local customs such as widow burning and female infanticide.

Early Buddhism offered a progressive approach to gender for its time. The Buddha allowed women to participate in monastic rituals and taught that all people have equal spiritual worth. In the 2,600 years that followed, however, gender parity in Buddhism has varied widely by culture. Similarly, the Abrahamic traditions have held long-standing tensions between their sacred text, theology, spiritual practices, and surrounding cultural mores in relation to gender. Abrahamic sacred texts, such as the Torah, Christian scriptures, and Qu'ran often instructed women to be subservient, and there are passages that have been interpreted to condemn same-sex relationships.

While the roots of a religion can have a profound effect on its adherents, the social context in which it is practiced today may impact followers even more. Despite Sikhism's emphasis on a genderless soul, for example, homosexuality remains controversial within the religion, with some leaders condemning it and others coming out in support of legal protections for same-sex marriage.

Other major world religions continue to have similar ongoing debates about the role of women and LGBTQ people in the life of the community. While mainstream Catholicism remains strictly divided into male priests and female nuns, most mainstream Protestant denominations permit women to become pastors, bishops, and other authorities in the church. Some Christian churches allow their leaders to officiate same-sex marriages, while others have banned these unions from church grounds. Within Judaism, there is also diversity in approaches, with more progressive temples ordaining women and LGBTQ people and supporting same-sex marriage while conservative branches see these actions as serious transgressions.

There are not many statements about religion with which most people would agree, except that religion is an extremely powerful force that influences and is influenced by culture. Because of this, different religions' approaches to gender and sexuality can have a large impact on how gender roles are defined in a society and on the success or failure of movements for gender equality.

What is the history of gender equality under the law in the United States?

From the beginning of recorded history until just a few decades ago, the laws of most western cultures and nations have assumed that the biological sex of an individual and that person's gender are congruent. The law viewed sex as strictly binary—an individual, at birth, was viewed as either male or female, depending on the appearance of that individual's outward genitals. From birth until death, a person's legal status was dependent on a determination made at birth.

Most cultures attempted to enforce gender expression and gender roles on individuals based on that determination. In modern western society, both men and women were expected to express their gender by the clothing they wore, the way they groomed themselves, and the way they behaved or conducted themselves.

Men were expected to be self-confident, bold, independent, intelligent, assertive, and even aggressive. They were discouraged from expressing sensitive emotions. Men's roles included work in all sorts of professions and jobs and participation in government. Women, on the other hand, were expected to be polite, accommodating, dependent, emotional, and nurturing. Women's roles included taking care of children and the home. Women were permitted to participate in very few types of work—midwifery and selling food at markets being some of the few acceptable positions. Even in religious settings, men were promoted to higher positions than women.

Cultures attempted to enforce gender roles—and usually even gender expression—regardless of the individual's own gender identity or desire not to be bound by the culture's expectation of them. Sometimes enforcement was by means of laws, such as those that allowed only men to vote, and sometimes enforcement was by other means, such as community pressure.

In the early American colonies, colonial governments imported English common law, which placed women in a separate class of persons from men in many respects. This different treatment was based on assumptions that women were physically, intellectually, and emotionally inferior to men. However, because women gave birth to children and children were a valuable commodity, women were viewed as requiring protection so that they could produce heirs.

In the colonies, and then in the states, following English common law, the legal status of free women depended on marital status. Unmarried women had many of the legal rights of men, such as the right to enter into contracts, to sue, to obtain and hold property, to live where they chose, and to receive wages for engagement in whatever occupations would accept them (usually not professions such as attorneys, doctors, and higher level teachers). Marriage changed a woman's status dramatically—women were viewed as legally nonexistent in a status called *coverture*, which suspended almost all individual rights, giving them instead to her husband.

The husband was viewed as the head of household. He controlled all assets and made all important decisions. If he wanted to move, his wife had to move with him or she was guilty of abandonment. Unless she came from a wealthy family that could make special legal provisions for her, the wife's property became the husband's, including any wages she might earn. He could do with that property whatever he wanted, no matter how reckless or ill-advised his decisions might be. The law granted the wife certain rights to any land or buildings she may have brought into the marriage, but that would not help a woman from a family that lacked real estate.

After U.S. independence, some of the strict rules governing married women's rights were loosened, but most remained. During the 19th century, as the nation became industrialized, women demanded changes in the legal rules governing the rights of married women, especially in terms of property and custody of children. Gradually, states passed laws enabling women to own property, keep their own wages, and enter into certain professions that had been barred to them. States also gave mothers more rights than they had in England, although fathers were still favored.

Women in the 19th century hoped for more than that. They began organizing for voting rights. The Civil War intervened, but afterward they hoped that they, along with formerly enslaved citizens, would get the vote. However, their hopes were dashed when the word *male* appeared in the Constitution for the first time in Section 2 of the 14th Amendment, indicating that women—both Black and white—were not guaranteed the right to vote in federal elections. Women received the vote in many states starting in the late 1800s, and then finally, in 1920, throughout the whole country.

Interestingly, Section 1 of the 14th Amendment of the Constitution, the same amendment that locked women out of the vote for 50 years, guaranteed to all citizens the "privileges and immunities of citizens of the United States" and to all persons (citizens or not) the "equal protection of the laws." However, when women in the 19th and first half of the 20th centuries brought their claims under Section 1 to the courts, they were systematically denied.

Finally, in 1971, in *Reed v. Reed*, the Supreme Court, faced with a blatantly discriminatory state statute, used the Equal Protection clause to rule for equality. In Reed, the court struck down a state statute that prioritized men over women as estate administrators. Subsequent decisions ended statutes that granted female but not male spouses of military personnel automatic benefits (*Frontiero v. Richardson*, 411 U.S. 677, 1973), required pregnant women to take unpaid leave after their first trimester (*Cleveland Board of Education v. LaFleur*, 414 U.S. 632, 1974), made jury service mandatory for men but voluntary for women (*Taylor v. Louisiana*, 419 US. 522, 1975), set the age of majority at 21 for men and only 18 for women (*Stanton v. Stanton*, 421 U.S. 7, 1975), allowed women to buy beer at a younger age than men (*Craig v. Boren*, 429 U.S. 190, 1976), and created distinctions in Social Security survivor benefits between men and women (*Califano v. Goldfarb*, 430 U.S. 199, 1977).

Most of these statutes were invalidated based on the idea that they were rooted in stereotypes about men or women. However, when a statute was based on a medical condition only women could endure, because of the differences between female and male reproductive organs, women did not fare as well. In *Geduldig v. Aiello* (417 U.S. 484, 1974), women argued that since only women can become pregnant, a California disability program that denied benefits for disabilities resulting from pregnancy but granted benefits for disabilities resulting from all medical conditions that could be suffered by men was unconstitutional and should be overturned. The Court refused, stating that the program did not discriminate on the basis of sex but distinguished between pregnant and nonpregnant persons. Similarly, even when pregnancy discrimination was challenged under Title VII of the Civil Rights Act, the Court held that pregnancy discrimination is not sex discrimination under Title VII (*General Electric Co. v. Gilbert*, 429 U.S. 125,

1976). Finally, Congress overrode this decision through passage of the Pregnancy Discrimination Act of 1978.

Women have fared better under Title VII of the 1964 Civil Rights Law (employment) and other federal statutes that prohibit discrimination based on sex when the challenged statute is not based on something that has to do with reproduction. For example, in *Los Angeles Department of Water and Power v. Manhart* (435 U.S. 702, 1978), the Court overturned a statute requiring female workers to make larger pension fund contributions than their male counterparts. That statute was based on statistics showing that women generally live longer than men.

Attorneys hoping to strike down laws restricting abortion or other forms of birth control have primarily argued that they are unconstitutional based on the "right to privacy" rather than because they represent sex discrimination. Many antiabortion laws have been upheld even though, as Ruth Bader Ginsburg stated during her confirmation hearing, "Abortion prohibition by the State . . . controls women and denies them full autonomy and full equality with men."

Sexual harassment in the workplace has become recognized as violating Title VII's prohibition against sex discrimination in employment. In *Meritor Savings Bank v. Vinson* (477 U.S. 57, 1986), sexual harassment was directed by a male against a female, which is the most common sexual harassment situation. In later cases, however, the sexual harassment was directed by a female against a male, a male against another male, or a female against another female. Thus, the question arose whether those cases of sexual harassment were also violative of Title VII. In other words, does Title VII cover same-sex harassing behavior that would violate Title VII if the behavior was between two people of different sexes? The Supreme Court said yes in the unanimous decision in *Oncale v. Sundowner Offshore Services, Inc.* (523 U.S. 75, 1998).

In June of 2020, after decades of advocacy, the Supreme Court decided that the prohibition on sex discrimination in

Title VII of the Civil Rights Act protected LGBTQ workers. The Court had heard arguments in *Bostock v. Clayton Co.*, which encompassed two lawsuits from gay men who claimed they were fired because of their sexual orientation, and *Harris Funeral Homes v. EEOC*, which involved a transgender woman, Aimee Stephens, whose employer fired her for her gender expression when she said that she would be dressing as a woman and otherwise presenting as a woman on the job. "An employer who fires an individual merely for being gay or transgender defies the law," Justice Neil M. Gorsuch wrote for the majority in the 6-to-3 ruling. "It is impossible, to discriminate against a person for being homosexual or transgender without discriminating against that individual based on sex."

What is gender-based violence?

Unequal power relationships between men and women have led, across continents and centuries, to an overwhelming level of violence against women, both in personal relationships and on a larger scale, within societies and as a weapon of war between groups.

On a personal level, intimate partner violence affects all communities, regardless of race, socioeconomic status, or religion. According to the National Coalition Against Domestic Violence, one in three women experiences domestic violence in her lifetime and there are over 20,000 calls placed to domestic violence hotlines in the United States each day. Half of female homicide victims are killed by their partners.

Intimate partner violence is not only physical or sexual but also verbal and emotional. Abusers may use tactics such as threats, name-calling, isolation from friends and relatives, financial withholding, and custody battles over children to exert control.

Intimate partner violence is not limited to cisgender men abusing cisgender women. The reverse is less common but often underreported. Same-gender relationships can also be affected by intimate partner violence. Queer and trans people who are abusive toward queer and trans partners can use specific strategies to intimidate and isolate. If one partner is "out," for instance, that person may threaten to out the other person if their demands are not met.

Sexual violence often plays a role in intimate partner violence. While we are taught to be scared of walking down empty streets at night for fear of rapists, most sexual assaults happen between people who know each other. It was not until the 1970s that the first laws were passed making it a crime for a husband to rape his wife.

In the United States, one in five women reports a lifetime history of rape. Women are victims more often than men, although 1 in 71 men endorse a history of sexual assault. Men are even more likely than women to underreport rape, and many carry shame and guilt related to feelings of being emasculated. Trans people, especially trans women of color, are disproportionately targets of sexualized violence and murder. Children of all genders are also among the most vulnerable. Twenty percent of adult women and 5 to 10% of adult men recall a history of childhood sexual abuse.

While victims of sexual violence fall into many categories, perpetrators of sexual violence are predominantly cisgender men. This is an overgeneralization, and cisgender women as well as transgender people do sometimes sexually victimize others, but the vast majority of incidents of sexual assault are perpetrated by cisgender men. In cases of child sexual abuse, women are the perpetrators in 14% of incidents involving male children and 6% of incidents involving female children. When women are the perpetrators, victims may be even more reluctant to come forward, fearing they will not be believed.

In 2017, the #MeToo movement sprang up following sexual misconduct allegations against film producer Harvey Weinstein, shedding new light on the prevalence of gender-based violence and victim-shaming. For the first time in history, women's stories were taken seriously, even when the accused man was in a position of power. This was in stark contrast to the responses only one year earlier to the revelation that soon-to-be U.S. President Trump had boasted about women that he liked to "grab 'em by the pussy." The #MeToo movement was not enough by itself to cause the cultural shift needed to prevent the confirmation of Brett Kavanaugh to the U.S. Supreme Court in 2018. Echoing reactions to Anita Hill's testimony against Clarence Thomas over 25 years earlier, Kavanaugh's accuser, Christine Blasey Ford, had her reputation dragged through the mud while Kavanaugh went on to a seat on the highest court in the nation.

One-on-one attacks are not the only way that gender-based violence is used to control groups of people. Among the most heinous of crimes is the use of rape as a systematic weapon of war. During times of conflict, sexual assaults often skyrocket due to rising tension and anger as well as newfound power of one group over another. However, there is also evidence that in many world conflicts leaders specifically instruct soldiers to rape to demoralize the other side. Recent wars in Vietnam, Bosnia, Rwanda, and Syria have resulted in serious psychological trauma in large populations of women targeted through sexual assaults. Those who are raped during war are often ostracized by their families, and many bear the burden of giving birth to a rapist's child.

Men have also been targeted through systematic sexual violence during wars, where men and boys are sexually assaulted and tortured. In some conflicts, men are given a choice between raping their female relatives or friends and being killed themselves. U.S. soldiers were found guilty of multiple instances of sexual assault, including sodomy and other forced sexual acts perpetrated against mostly male detainees, during the Iraq war at the military prison Abu Ghraib.

Gender-based violence sometimes receives press coverage when it occurs during armed conflicts, but is often overlooked when used to control populations during times of relative peace. The techniques employed during these times may not be as sensationalistic, but affect the lives of women and girls on a daily basis. Child marriage, for instance, has many lasting effects on women, including poor educational outcomes and high rates of maternal mortality. Genital mutilation is another systematic strategy for limiting women's ability to engage fully in the world, taking away their potential for pleasure, to control their sexuality, with risk of infections and even death.

There are also more subtle, but pervasive ways in which gender-based violence keeps women from reaching their full potential. Even women who deny ever being victims of gender-based violence can recount numerous life experiences in which they felt threatened. Almost no woman in the world makes it to middle age without being on the receiving end of unwanted sexual comments or touches. Because of the always-present threat of violence against women that exists in most societies, women spend much of their mental energy every day thinking about how to stay safe. They avoid going out at certain times of day. They stay away from particular places, even within their own towns or cities. They steer clear of people who they know are potentially dangerous. They often live circumscribed lives to lower their chances of being victimized. All of this takes a toll, robbing even women who are not victims of their freedom to exist in the world in peace.

What are the legal protections for transgender people in the United States?

Until the Supreme Court's 2020 decision in favor of the rights of LGBTQ employees, in most areas of the United States, there were no explicit laws preventing a person from being fired simply based on their gender identity. Today transgender people can be refused entrance to restaurants and hotels. They can be barred from military service. They can be turned away from adoption agencies and have their healthcare needs denied coverage by their insurance companies. Despite improvement in social recognition of trans people, our legal system lags behind.

Although some cities and states put into place protections for trans employees, there was no federal law in the U.S. barring discrimination in employment based on gender identity until June of 2020, and, at the time of publication, there is not yet clarity on how the Supreme Court's decision will be upheld. The federal Employment Nondiscrimination Act was originally brought to Congress in 1994 to protect against discrimination based on sexual orientation. In 2007, gender identity was added, causing controversy, as some well-known LGBTQ organizations, including the Human Rights Campaign supported the bill without gender identity, not believing it would pass with it. The bill, even without gender identity, passed the House but not the Senate. A trans-inclusive version of the bill was been brought up in subsequent congressional sessions, but time and time again failed to move forward.

Another approach lawyers have taken to gaining employment protections for transgender people is through interpretation of existing laws. Title VII of the Civil Rights Act of 1964 prohibits discrimination in employment based on sex, race, color, national origin, or religion. In some recent court cases, Title VII has been interpreted to apply to transgender people through the "sex" category because trans people fail to conform to societal expectations for their sex. This was the approach taken by lawyers in the most recent Supreme Court case on this matter, which was decided in June 2020 in favor of LGBTQ workers.

The Civil Rights movement of the 1960s has also had a delayed positive effect on transgender people's housing security. For many years, including well into the 2000s, LGBTQ people had little recourse if a landlord decided not to rent to them. However, recently there have been successful court cases arguing that the Fair Housing Act, part of the Civil Rights Act of 1968, which prohibits property owners from discriminating based on sex in the sale or rent of their buildings, applies to gender identity and sexuality. In public housing, there are even more straightforward protections. The Department of Housing and Urban Development (HUD), which oversees shelters, subsidized housing projects, federal housing vouchers, and federally insured mortgages, implemented the Equal Access Rule in 2012, prohibiting discrimination based on sexual orientation or gender identity. Unfortunately, HUD regulations are not legislative decisions and can therefore be updated at any time by the secretary of HUD, who is appointed by the president.

Unlike Title VII, Title II of the Civil Rights Act of 1964, which covers discrimination in public accommodations such as hotels, restaurants, and entertainment venues, includes only race, color, religion, and national origin, but not sex, in its list of protected classes. This leaves whole categories of people—women, men, transgender people, and gay, lesbian, and bisexual people—open to discrimination in access to these types of facilities. There are some states, such as California, that have enacted their own legislation to protect against discrimination in public accommodations, but the majority of states offer no such protection and permit restaurant, bar, and hotel owners to deny entrance to LGBTQ people.

Within their own homes, transgender people also face significant legal barriers. The country-wide legalization of same-sex marriage in 2015 was a step forward not only for same-sex couples but also for transgender people in different-gender relationships because they were now able to marry even if their genders were not legally recognized. However, unlike cisgender heterosexual couples, same-sex couples and those with a trans partner continue to face uphill battles to legitimize their relationships, as well as their connections to their children. In most cases, it is necessary to go through the process of a second parent adoption, meaning that the partner who did not give birth to the child formally adopts their own child to ensure parental rights. Even with a second parent adoption, when a couple with a child splits up, the trans partner often faces discrimination in the process of seeking legal rights to their child. Decisions about custody and visitation are made by individual judges on the basis of the "best interests of the child," leaving trans people open to judicial bias.

Same-sex marriage has also has positive effects on trans people's immigration rights, allowing recognition of all married couples in the immigration process, regardless of the genders of the partners. Still, trans people face numerous barriers to legal immigration. They are often encouraged not to change their legal names or genders during the application process because any glitch can cause delays, which results in many people putting off important changes while they wait for long periods for their paperwork to go through. Additionally, while trans people can apply for asylum as part of a persecuted social group, similar to family courts, immigration courts are presided over by judges who have the final say in whether asylum will be granted, and judges can display prejudice against trans communities. While waiting for their court appearances, some asylum seekers and other immigrants are held in detention centers run by Immigrations and Customs Enforcement (ICE). These facilities are notoriously dangerous for vulnerable groups. In 2018, a trans woman from Honduras named Roxsana Hernandez Rodriguez died in ICE custody and her autopsy showed signs of abuse.

Violence against trans people does not occur only in ICE detention centers, but also in many different settings across our communities. A 2013 study by the National Coalition of Anti-Violence programs showed that 72% of hate crime murder victims were trans women, and almost all of those were women of color. In 2009, federal hate crimes laws were expanded to cover sexual orientation and gender identity. Whether these laws are effective in preventing crimes, however, is debatable, and some trans lawyers and activists argue that hate crimes laws simply expand prison populations rather than fixing underlying problems.

Concerns over expanding prison populations are directly relevant to trans people, as 16% of trans people have been incarcerated compared to 3% of the general population. Trans youth grow up experiencing more trauma and have fewer options for school and employment, leading to increased rates of sex work and arrests for prostitution. Police often target trans people even when they are not engaging in sex work. Once in jail or prison, trans people are victimized by both staff and other inmates. Many trans women, especially, are put in extremely dangerous situations, where they are assigned to units with cisgender men. One solution staff turn to is separating out trans prisoners, but this is typically to solitary confinement, which has been proven to be psychologically damaging. Eighty-five percent of LGBTQ people who have been in prison report spending time in solitary confinement.

Despite there being an estimated 15,500 transgender active duty, national guard, and reserve personnel, as well as 134,300 trans veterans, the status of trans people in the military remains precarious. In 2016, under the Obama administration, Department of Defense regulations banning transgender service members were repealed. However, since the election of Donald Trump, there have been a number of twists and turns in the saga of trans people interested in or currently part of the U.S. military. Beginning with a July 2017 tweet in which Trump announced that transgender people would no longer be able to serve in the military, there have been numerous back-and-forth arguments, including a letter signed by 56 retired generals and admirals in favor of transgender troops. Lawsuits have continued to block implementation of Trump's plan, and the Pentagon is listening to the courts for now.

Engaging in any public activity, including looking for a job or apartment or joining the military, involves proper identification. For trans people, this means identification with the correct name and gender. Name changes can typically be accomplished relatively quickly, although the process is not always smooth. Most states require a court order, which involves appearing in front of a judge. Most people are permitted to change their names as they please, but judges sometimes demonstrate discriminatory behavior toward trans applicants. In addition to going in front of a judge, there is also often a requirement to publish the name change in a local newspaper. This stipulation exists to prevent name changes for the purposes of escaping debt or legal issues. For trans applicants, this can pose a safety risk and the requirement is sometimes waived.

Gender changes are more complicated than name changes and depend on the level (federal or state) at which the change is being requested. After years of obstructive policies, in 2010, the State Department announced a new policy that gender could be changed on federal passports and social security documents with a letter from a doctor indicating the person had "appropriate clinical treatment for transition to the updated gender." Because this phrase allows physicians to make their own judgements about what is appropriate treatment, those trans people with access to supportive physicians are able to apply for federal gender changes relatively easily, while those who do not have this kind of access often have to jump through significant hoops.

State-level gender changes on ID documents such as drivers licenses and nondriver IDs vary by state. Alabama, for instance, requires a letter written by a surgeon stating that gender confirmation surgery has been completed, effectively barring anyone who cannot afford or does not desire surgery from changing their gender marker on their ID. There has been significant progress in many states, however, and in 2017, Washington, D.C.; Oregon; and California even began to offer a gender-neutral option on drivers licenses. Gender changes to birth certificates are also processed on a state level, and depend on the state in which a person was born. Most states will reissue or amend the gender on a birth certificate. Some require genital surgery. The original birth certificate may or may not be "sealed" and the amendment may be obvious, making it difficult to keep the change confidential. Four states (Idaho, Kansas, Ohio, Tennessee) will not reissue or amend birth certificates for gender changes under any circumstances, leaving transgender people there at risk of discrimination in any situation in which they are required to present their birth certificates.

6

CONTEXTUALIZING GENDER

What are the different "waves" of feminism and how are they distinguished?

In March 1968, *The New York Times Magazine* published an article by Martha Weinman Lear titled, "The Second Feminist Wave," exploring the activities of women's groups such as the newly founded National Organization for Women. Lear's article may have been the first to introduce to the public the idea of feminist "waves."

Today, we talk about the three (or four, or five, depending on who you ask) waves of feminism, for the most part accepting them as fact. The first wave is usually described as lasting from the Seneca Falls Convention in 1848 until women's suffrage in 1920; the second, as the 1960s and 1970s; and the third, as beginning in the 1980s or 1990s and lasting until today, or perhaps making way for a fourth or even fifth feminist movement in the last decade or two.

The impetus to divide history into discrete periods is understandable. It helps us to break down complicated events into more digestible chunks. However, it can also be limiting, preventing us from seeing the whole picture. The concept of a "wave" suggests that most of the work of that period took place within a discrete interval. The first wave of feminism, for example, is said to have ended in 1920, and the second wave to have begun in the 1960s. Does that mean that women were not organizing to improve their lives for the 40 years in between?

The term *wave* also suggests that a group of people is working in concert with each other and in agreement about what they are fighting for and how the fight should occur. However, when we look back on feminist movements or consider feminist work being done today, it is clear that conflicts about goals and approaches have always been and continue to play a fundamental role. In addition, movements do not grow in isolation—the first wave of feminism coincided with abolitionist causes, and the second wave, with the Civil Rights movement. Each influenced and interconnected with the other.

The year 1848 is thought of as the official beginning of the first wave of feminism in the United States because it is when the Seneca Falls Convention occurred. In the traditional story about the origins of Seneca Falls, the young abolitionist Elizabeth Cady Stanton was so devoted to ending slavery that she chose to spend her honeymoon at the World Anti-Slavery Convention in London. According to legend, she was not yet fully aware of her own subjugation as a woman until, arriving at the convention, she and other women were told that they could not participate and would have to sit in a cordoned-off section in the back without speaking. As the story goes, this is where Stanton met Lucretia Mott, and the two decided that something had to be done about women's rights, leading to plans for Seneca Falls, the first ever American women's rights convention.

While there is truth to this story, it is also somewhat mythologized. Elizabeth Cady Stanton and Lucretia Mott did meet at the World Anti-Slavery Convention. However, that convention occurred eight years prior to Seneca Falls, in 1840, and there was no direct link between the two events. In fact, Stanton gave birth to her first of seven children in 1842 and spent much of her time in the years between the two conventions preoccupied with the task of raising her children and not planning for Seneca Falls.

Additionally, the World Anti-Slavery Convention was in no way the first time Stanton or Mott had been faced with the idea that women were treated unfairly. Living their daily lives, they encountered sexism all around them, and they were both aware of work being done by other women related to women's rights. In 1819, Emma Willard had presented her "Plan for Improving Female Education" to the New York Legislature. In 1828, Sojourner Truth, having escaped from slavery, sued her former master for custody of her son and won the case. In the 1830s, the abolitionist Grimke sisters traveled the country speaking to mixed-gender audiences, flying in the face of what was considered proper for women at the time.

There are many myths surrounding the Seneca Falls Convention itself, the most important of which is that it was organized around the push for women's suffrage. The focus of the Convention was the creation of a Declaration of Sentiments, based on the Declaration of Independence, arguing that "all men *and women* are created equal." The declaration included language on topics such as marriage and divorce, property rights (married women at the time could not own property), and employment (many careers were closed to women). It was considered radical when Stanton introduced a resolution stating that women should have the right to vote. In fact, Mott and many others opposed the idea. Although we are taught in school that the abolition and women's suffrage movements were separate, fascinatingly, it was the ardent feminist Frederick Douglass, the only Black person present at the convention, who stood up and addressed the crowd on the topic of suffrage, arguing that without the vote for women, the country would miss out on "one half of the moral and intellectual power" of the world.

The connections between abolition and women's suffrage would become even more evident later in the 19th century, as the Civil War approached. Women's rights leaders such as Stanton and Susan B. Anthony worked closely with Frederick Douglass and others to build a case for universal suffrage—the idea that all adults, regardless of race or sex, should have the right to vote. However, when the Civil War ended and constitutional amendments (called the "reconstruction amendments") were moving through Congress, it became clear that Black men were likely to be granted the vote, but not women.

Friends who had once stood united, such as Susan B. Anthony and Frederick Douglass, were now torn apart, each arguing for different sides. Sojourner Truth, decades before, in 1851, had given her famous "Ain't I a Woman" speech, pointing out the differences in the way Black and white women were treated. After the Civil War, as one of the few Black women with access to high-level suffrage meetings, she argued that "if colored men get their rights, and not colored women theirs, you see the colored men will be masters over the women, and it will be just as bad as it was before." Opposing her, Frederick Douglass, despite having stood up for women's rights for most of his life, made an impassioned speech asking for others to consider the dire needs of Black people. "When women, because they are women," he said, "are dragged from their homes and hung upon lampposts, when their children are torn from their arms and their brains dashed upon the pavement . . . then they will have the urgency to obtain the ballot."

Suffrage did, in fact, pass for Black men and not women, creating chasms between activists and leading to a racist backlash from some of the feminists, such as Anthony and Stanton, who we consider our national heroes. This dark chapter in our history often does not make it into school books.

Given that they were already adults by the time of the Seneca Falls Convention, many of the women involved in the early suffrage movement were quite old by the end of the 19th century, and almost all became too frail to continue to participate or passed away by the dawn of the 20th century. There was a period of relative quiet at the end of the 1800s where there seemed to be little forward momentum. While the first "wave" of feminism is considered to be from 1848 to 1920, it might be more accurate to describe the first wave as two separate periods. Four states granted women the right to vote between 1890 and 1896, but it was not until 1910 that women in other states began to be successful in their state suffrage bids. Then, first wave feminism "part two," equipped with new leaders, including Alice Paul and Carrie Chapman Catt, brought the final push needed to secure the 19th Amendment, granting nationwide women's suffrage, in 1920.

Though the first wave of feminism is said to have ended in 1920, and the second wave to have begun in the 1960s, there were significant changes in women's lives in America in the interim. World War II, especially, brought many women into the workforce. Planned Parenthood was founded in 1942. The 1950s showed a return of a strictly gendered culture, but this did not last long. Simone DeBeauvoir's *Second Sex* was published in the U.S. in 1953. In 1960, the first birth control pill was approved by the Food and Drug Administration. Women observed and participated in the Civil Rights movement of the late 1950s and early 1960s, honing techniques, such as acts of civil disobedience, that they would later use in feminist work. Ultimately, the "in-between" period of the 1920s to the late 1960s led directly to the more publicly feminist era of the 1960s and 70s.

The second wave of feminism was one of the most important movements in the history of the United States, and, like the first wave, is difficult to condense into a short description. During the 1960s and 70s women achieved massive successes, including abortion rights and sexual harassment laws. Title IX, passed as part of the Education Amendments of 1972, outlawed discrimination in educational programs, having a profound impact on women's lives and careers. From a social perspective, women made inroads into male-exclusive establishments and challenged traditional understandings of heterosexual relationships.

Some historians have described the second wave as two separate movements with different goals. These types of divisions were not unique to the second wave. Suffragettes in the early 20th century had taken opposing stances, some going on hunger strikes and throwing bricks through windows and others attempting to make change more diplomatically. Second wave feminists have been divided into "equal rights" and "radical" feminists, the former involved in changing the social and legal landscape to give women the same rights as men, and the latter hoping to disrupt the foundations of existing systems, such as patriarchy and capitalism, rather than working within them. Though there were certainly different approaches and instances of bitter disputes, there were also many women who held more radical views but compromised to ensure smaller gains along the way.

In addition to the differences in opinion about strategic approaches to feminist work in the 1960s and 70s, there were also issues of race, class, and sexuality that permeated the feminist movement. Many of the movement's leaders were upper middle class straight white women, and the experiences of women of color, working class women, and lesbians were often pushed to the side. Women like Angela Davis, Florynce Kennedy, Alice Walker, and many others helped to create a vast literature of Black feminist thought. In 1969, when Betty Friedan, president of the National Organization for Women, described the threat of lesbian involvement in the women's movement as a "lavender menace," lesbians responded by protesting at the following year's Congress to Unite Women, in which there were no openly lesbian-identified women on the program. These challenges to mainstream second wave feminism would pave the way for new approaches as feminist waves moved forward.

The third wave of feminism is said to have begun in the 1990s, though it had roots in the 1980s, a time most consider to be an era of conservative backlash. Still, it was in the late 1980s that Kimberle Crenshaw introduced the concept of intersectionality, which would become one of the centerpieces of the third wave feminist movement. Intersectionality emphasized the multiple systems of oppression that work together to influence our lives. Many third wave feminists saw the movement as a response to more monolithic representations of women and their needs during the second wave.

The riot grrrl scene, which began in the pacific northwest in the early 1990s, remains emblematic of certain subcultures of third wave feminism. Bands and fans called out sexism, sexual assault, and homophobia in their songs and zines (photocopied homemade magazines). The language they used ("girl" rather than woman) and their clothing aesthetic (femme-punk style) represented a belief that second wave feminists had downplayed and sometimes even looked down on femininity to fight for equality. Instead, riot grrrl third-wavers embraced femininity and girliness, arguing that they could be both feminine and strong.

Support of transgender people improved with the arrival of third wave feminism. Some, but not all, second wave feminists subscribed to an "essentialist" idea of what it meant to be a woman (i.e., a person born with a vagina). A few even went as far as to claim that trans women were men attempting to "infiltrate" and take over women's spaces. While there were still anti-trans feminists involved in third wave movements, they were not as common. Interestingly, it has been subsets of self-identified "radical" feminists—once considered on the most cutting-edge—who have continued to hold the most bigoted beliefs about trans people. Not all radical feminists are anti-trans, but those who are have been referred to as trans-exclusionary radical feminists, or TERFs.

It is not completely clear if feminist movements today are part of a continued third wave or a new fourth wave. Some argue that a fourth wave began between 2005 and 2010, with the introduction of social media platforms such as Facebook, Twitter, and YouTube. These allowed not just feminist movements, but many other campaigns and causes to spread their messages quickly and easily, resulting in uprisings like the Arab Spring. Hashtags, including #metoo, #timesup, and

#yesallwomen created solidarity in new ways. Social media also allowed for mass organization of events like the 2017 Women's March on Washington. In many ways, it does not matter how we divide feminist "waves," or which one we are in right now. What does matter is that feminist movements continue to grow and learn from the past. Today's feminist movements are more intersectional than they ever have been. And tomorrow's will be even better.

What is intersectionality?

"Ain't I a woman?" asked Sojourner Truth to her mostly white, upper class audience at the Women's Convention in Akron, Ohio, in 1851. "That man over there says that women need to be helped into carriages, and lifted over ditches, and to have the best place everywhere. Nobody ever helps me into carriages, or over mud-puddles, or gives me any best place! And ain't I a woman?"

When Sojourner Truth gave her famous "Ain't I a Woman?" speech, she was pointing out the differences in the way Black women were treated compared to white women and challenging the notion that white women's experiences were generalizable.

It may come as a surprise—because women of color have been making this point for so long—that academics did not have a word for the concept of *intersectionality* until it was introduced by lawyer and feminist theorist Kimberlé Crenshaw in a paper on feminism and antiracist politics in 1989. Crenshaw argued that the experience of being a Black woman could not be explained simply by discussing the experiences of being Black and being a woman, but instead represented a product of interactions between the two.

As an example of the need for intersectionality, Crenshaw drew on the case of a group of Black women who sued General Motors (GM) in 1976, alleging employment discrimination. Because the court did not understand the concept of intersectionality, GM's lawyers were able to claim that the company was not biased by presenting evidence that it hired both Black men and white women. The court stated that according to the Civil Rights Act of 1964, it could not consider a case of combined race- and gender-based discrimination.

Today, intersectionality is employed both within feminism and other systems of thought as an approach that takes into account the intersecting systems of power and oppression, such as race, class, gender, age, ability, and sexual orientation that work together to determine our experiences.

Inherent in the concept of intersectionality is the idea that one person may experience privilege related to some of their identities and oppression due to others. A wealthy Black cisgender gay man, for instance, may benefit from his status as cisgender, male, and wealthy, but be discriminated against due to his race and sexuality. Each person's intersecting identities determine their position in the world and are complicated and unique.

What is misogyny?

In simplest terms, misogyny can be defined as hatred or contempt for women. The word was formed from the Greek roots *misein* (to hate) and *gynē* (woman). Patriarchal societies across time and around the world have typically incorporated misogyny into their cultural beliefs, often with devastating consequences on the psychological well-being of both men and women. Girls who grow up in misogynist societies can experience internalized misogyny as well as learn to perpetrate misogyny toward other women. Boys who grow up in misogynist societies can do serious harm to the women in their lives as well as to themselves and other men by being overly critical of any traits or behavior perceived as feminine.

Many would argue that misogyny has decreased significantly in modern western societies. However, in today's world, online misogyny has become increasingly more vicious as social media and internet forums have developed into everyday tools. One of the most well-known cases of severe online misogynistic harassment occurred as part of what is referred to as the Gamergate controversy. In 2013, video game developer Zoe Quinn created a game called Depression Quest, centered around her experiences with depression. Soon after, her ex-boyfriend, Eron Gjoni, wrote a blog post disparaging Quinn's work and accusing her of cheating on him with multiple men, including one who Gjoni claimed she slept with to get a good review of her game. Using the hashtag #gamergate, thousands of men piled on, taking the moment as an opportunity to threaten and belittle Quinn, who was known for speaking out about gender inequities in gaming. Quinn received rape and death threats, was encouraged to commit suicide, and was doxed—her home address and personal information released to the public. Many other women were affected by #gamergate, including Anita Sarkeesian, who was a vocal critic of traditional women's roles in video games. Sarkeesian received numerous death threats, forcing her to cancel events, and one harasser even created a video game that allowed players to beat up a character modeled after her.

Another arena where misogyny has come front and center is with the news coverage of "incels," or those who call themselves "involuntarily celibate," and blame their lack of sexual experience on women, who they feel are inferior to men and should give them sex when they want it. In April 2018, a man named Alek Minassian drove a van through a crowded business district in Toronto, killing 10 people and injuring 16 others. Minassian identified as an incel, as did Elliot Rodger, who killed 6 people and injured 14 in a stabbing and shooting rampage near the University of California–Santa Barbara in 2014. Prior to his attacks, Rodger wrote a 107,000-word manifesto in which he argued for a "War on Women" for "depriving me of sex."

Unfortunately, misogyny is ingrained in our societies in deeply troubling ways, and will continue to proliferate until we have done the work of changing the underlying belief that women are inferior to men.

What is "toxic masculinity"?

Toxic masculinity has been the topic of much discussion in recent years. The Good Men Project defines toxic masculinity as follows:

> Toxic masculinity is a narrow and repressive description of manhood, designating manhood as defined by violence, sex, status and aggression. It's the cultural ideal of manliness, where strength is everything while emotions are a weakness; where sex and brutality are yardsticks by which men are measured, while supposedly "feminine" traits—which can range from emotional vulnerability to simply not being hypersexual—are the means by which your status as "man" can be taken away.

Masculinity itself is not inherently negative or harmful in nature. Healthy masculinity embraces traits that feel comfortable to those who see themselves as masculine without causing damage to themselves or others. Healthy masculinity leaves room for nurturing, communication, and vulnerability. Toxic masculinity, on the other hand, is the result of unhealthy cultural expectations of men and manhood and reduces the concept of "being a man" to exerting physical power and dominance.

The harmful effects of toxic masculinity are far reaching. Whereas women, girls, and gender-diverse people such as transgender and nonbinary individuals suffer devastating and often lethal byproducts of toxic masculinity, cisgender men are also heavily impacted. After all, if one is raised in a culture in which doing your gender "right" lies in your alignment with an expectation of being oppressive, dominant, forceful and unemotional, disengaging can easily be interpreted as being weak. Not living up to social expectations of manhood can thrust one into being perceived as a "pussy" or "cuck," lacking balls and lacking manhood.

Toxic masculinity involves the need to aggressively compete with and dominate others. Scholars who explore the social construction of masculinities have warned of the psychological and sociological red flags this mindset engenders. Prison studies have explored the ways in which elements of toxic masculinity prevent rehabilitation, engagement with mental health systems, and family reunification. Researchers who focus on schools and education warn of the harmful byproducts of toxic masculinity on developing young minds, and a more recent focus on war and terrorism has come to a similar conclusion. Toxic masculinity has prevailed as a powerful weapon in the proliferation of violence between people, countries, and continents.

Psychiatrist Terry Kupers identifies toxic masculinity as "the constellation of socially regressive male traits that serve to foster domination, the devaluation of women, homophobia, and wanton violence." While there is nothing particularly toxic in a man's pride in his ability to win or succeed, Kupers warns that in settings that lack resources, safety, and supportive community, such as prison, these drives can easily become blinding.

What's more, toxic masculinity is bracketed within larger systems of privilege and power and can be layered with racialized dynamics, inhumane treatment, and poor service provision. Race and class, access to education and healthcare, ability to think critically (and without scarcity) about the future, and engagement with diverse male role models all impact the way masculinity is perceived and embodied. In a recent Huffpost article "More Men Should Learn The Difference Between Masculinity and Toxic Masculinity," author Ryan Douglas proposes,

> Toxic masculinity in American culture starts with straight, white men and trickles down through marginalized groups, affecting the way they perceive themselves and behave. We can't examine straight African-American men's behavior, for example, without first examining the white power structure that influenced it. And we can't separate how black men treat women from how white men treat black men.

In one simple statement, Douglas negates much of the pushback to this conversation: "Masculinity is real, natural, and biological. Toxic masculinity is a performance invented to reinforce it." The reinforcement of masculinity in western societies often includes high degrees of ruthless competition, an inability to express emotions other than anger, an unwillingness to admit weakness or dependency, devaluation of women and all feminine attributes in men, and homophobia.

Today, there are definable masculinities that present alternatives to the hegemonic ideal. Gay and transgender men often model masculinities that deviate from the toxicity prevalent throughout society, along with intellectuals, geeks, artists, musicians, and stay-at-home dads (none of these are exempt from adopting toxic masculinity but may present alternatives). Intentional self-evaluation is important for all people aligned with any type of masculinity. Australian author Tim Winton writes about his experience watching "the tenderness [being] shamed out of [the boys]" through years spent rehearsing and projecting masculinities, watching for subtle reinforcement and validation. What would it take to signal strength in empathy, listening, and shared space?

After all, toxic masculinity is a burden to men. Winton warns of the ways in which men too are "shackled" by misogyny. "It narrows their lives," Winton writes, "Distorts them. And that sort of damage radiates; it travels, just as trauma is embedded and travels and metastasizes in families. Slavery should have taught us that . . . [m]isogyny, like racism, is one of the great engines of intergenerational trauma."

Undoing toxic masculinity is about undoing the ways in which our culture commodifies power. Perhaps the most influential intervention to change an overwhelming climate of toxic masculinity would be first to create systems to aid in recognition of its patterns, identifying which specific dynamics are harmful and to whom.

What are Women's Studies, Men's Studies, and Gender Studies?

In 1956, feminist scholar Madge Dawson began teaching a class called "Woman in a Changing World," within the Department of Adult Education at Sydney University in Australia. The course focused on the socioeconomic and political status of women in western Europe and is known today as one of the first Women's Studies courses taught at the university level. A decade later, in 1969, the first accredited Women's Studies course was taught in the United States at Cornell University. San Diego State College, now San Diego State University, established the first Women's Studies program in 1970, and the first PhD program in Women's Studies began at Emory University in 1990. Today, Women's Studies courses and accredited degree programs exist throughout the world. In 2015, Kabul University in Afghanistan began their first master's degree course in gender and women's studies. Doctoral level Women's Studies programs exist today in Budapest, Japan, England, Australia, the United Kingdom, Wales, Canada, and the United States. Other degree granting programs currently operate in Ghana, Croatia, Poland, Slovakia, Sudan, Uganda, Palestine, Malaysia, Korea, and Barbados, and many other places.

The United States National Women's Studies Association was established in 1977. The association defines the field as having its roots in the student, Civil Rights, and women's movements of the 1960s and 1970s. The field of Women's Studies has expanded from its early origins and has looked to understand the role and impact of gender, and systems of gender, on university life and scholarship, structures of power and privilege, and sociopolitical patterns and trends. By incorporating feminist theory, social justice models, and queer theory, Women's Studies programs have worked to locate themselves within a changing and complex world.

In the last two decades, many Women's Studies programs at the university level have begun to use the term *Women's and Gender Studies* as a categorical shift in title and practice. Other offshoot programs have emerged under the descriptor *Gender Studies*. Gender Studies programs typically include the same theoretical approaches as Women's Studies, focusing in particular on taking an intersectional approach to personal identity and social dynamics. Gender Studies scholars propose that patriarchy functions to regulate people of all genders, not just women.

Gender Studies programs have largely emerged alongside a line of inquiry questioning the binary notion of gender and have centered around the ways in which the experiences of women and other gender minorities add to a more comprehensive understanding of gender. Gender Studies programs have therefore relied heavily on queer theory and the study of sexuality, understanding the inherent relatedness between gender identity, sex, sexuality, politic, and liberty. Whereas the field of Women's Studies has been associated with second wave feminism, Gender Studies emerged alongside third wave feminist discourse.

While, on one hand, the shift from Women's Studies to Gender Studies could signal a more inclusive course variety, accounting for the full range of gender diversity and experience, the shift has not always been welcomed. As the field has redefined itself to think more holistically about gender, the space devoted exclusively to women has changed (as has the definition of women), creating room for Masculinity Studies and Queer Studies.

Men's Studies, sometimes referred to as Masculinity Studies, is an interdisciplinary field devoted to the interrogation of masculinity through analysis of cultural, social, historical, political, psychological, economic, and artistic themes. Men's Studies programs are largely feminist in nature and revolve around conversations similar to those taking place in Women's Studies and Gender Studies classrooms, with somewhat more intense focus on the roles of men and anyone else who identifies as masculine. The American Men's Studies Association traces the roots of an organized field of Men's Studies to the early 1980s, although a small number of Men's Studies courses were taught as early as the 1970s. Content within Men's Studies classrooms is largely critical of the role of male privilege and misogyny, seeking to understand men not as inherent perpetrators, but rather as additional victims under the crushing weight of sexism.

What is queer theory?

The word *queer* entered into the English language in the 16th century and originally meant "strange" or "odd." Its use as a pejorative term to refer to gay and lesbian people began near the end of the 19th century. In the late 1980s, gay, lesbian, and bisexual people first reclaimed *queer* for themselves. Those who identified as queer meant to set themselves apart from mainstream gays and lesbians who they saw as accepting integration into the larger society without challenging underlying power structures. Queer people were anti-assimilationist and fought against more conservative gay groups that focused on same-sex marriage and military inclusion. One of the first organizations to use the word *queer* was Queer Nation, a nonhierarchical, decentralized, direct action group that was known for using confrontational tactics to protest violence against LGBTQ people.

At the same time that activists were beginning to use the word *queer* to label themselves and their activities, academics emerged who hoped to use the same term to radicalize gay and lesbian studies. Teresa de Lauretis, a professor at the University of California, Santa Cruz, is credited with introducing the phrase *queer theory* at a conference at the University of California-Santa Cruz, in 1990. From then on, queer theorists ventured into a diverse array of subjects, including LGBTQ history, social construction of identities, and literary theory.

In the few decades prior to the appearance of queer theory, a number of researchers had taken on the herculean task of tediously gathering evidence of same-sex relationships and gender nonconformity throughout history. The majority of these investigations pertained to male homosexuality. A notable exception is Lillian Faderman's 1985 book, *Surpassing the Love of Men*. Faderman argued that, contrary to popular belief, emotionally intimate (and sometimes sexual) relationships between women (known as "romantic friendships") were common and rarely pathologized during the 16th to 19th centuries. Gender transgressions (wearing male clothing, speaking in public, taking on male professions), however, were frowned upon. According to Faderman, the first wave of feminism (which included the fight for suffrage) coincided with the new "science" of sexology in the early 20th century, leading to a more intense focus on women's relationships with each other and to the pathologization of intimacy between women.

Queer theorists took a different approach to LGBTQ history. Rather than using a long history of same-sex relationships to demonstrate that gays and lesbians had existed throughout time, they instead proposed that, although same-sex relationships had certainly occurred, the categories "gay" and "lesbian" were new. For many scholars, queer theory has its roots in the work of French philosopher Michel Foucault, whose *History of Sexuality*, the first volume of which was published in 1976, made novel arguments about the state of sexuality over the last few centuries. According to Foucault, despite our belief that sexuality was extremely repressed over the 17th to 19th centuries, there is evidence that this period was a time of outward repression but simultaneous proliferation of literature and study of sexuality. Combing through historical texts, he concluded that same-sex relationships had occurred throughout time, but that until the late 19th century, people did not have sexual identities in a modern sense. Rather than a population being divided into straight and gay, anyone could participate in any sexual act. David Halperin made a similar argument in his 1990 book, *One Hundred Years of Homosexuality*, which looked at homosexuality in ancient Greece, a time and place when gay and lesbian identities had not yet been "invented." "The sodomite had been a temporary aberration," Foucault wrote. "The homosexual was now a species."

A number of reasons have been proposed for this shift from an understanding of sexuality as sexual acts to the newer concept of each person possessing a particular sexuality. One of the best known theories was put forth by writer John D'Emilio, who argued that industrialization in the 19th century led many people to leave their families behind to move to cities. Urban landscapes provided an environment in which workers were no longer tied to family units and could interact sexually with others of the same sex on a regular basis, creating communities and developing new identities.

Following in Foucault's footsteps, Judith Butler proposed that it was not only sexuality, but also gender that had been socially constructed. Earlier feminist scholars had written about this topic, including Gayle Rubin, who challenged "the idea that sex is a natural force that exists prior to social life" (p. 275). Going even further than Rubin, in her 1990 book, *Gender Trouble*, Butler argued that, contrary to essentialist feminist ideas of women as an unchanging category, there is no "real" or "true" gender before culture. According to Butler, gender is instead socially constructed through clothing, speech, and nonverbal communication that becomes so obfuscating that it makes us believe gender is an essential part of our being. Butler is known for advancing the concept of "performativity" of gender, meaning that we each act and dress in ways that continually reinforce gender. Butler offered a critique of "identity politics," in which groups fight for rights based on their identities because she felt that our social categories were created by humans and that we should be fighting for the human rights of all people rather than dividing into socially created categories.

Another well-known early queer theorist was Eve Kosofsky Sedgwick, an English professor whose 1985 book *Between Men: English Literature and Male Homosocial Desire* and 1990 book *Epistemology of the Closet* argued that an understanding of western culture is never complete without incorporating an analysis of homosexual relationships. She encouraged a generation of scholars to investigate queer interpretations of literature.

Following the early queer theorists, there was a surge in scholarship, with contributions from a wide variety of participants, including queer people of color and transgender people. These include Roderick Ferguson, who first introduced the term *queer of color critique* in his 2004 book *Aberrations in Black,* as well as Jack Halberstam, Patrick Califia, Samuel Delaney, Jose Munoz, Gayatri Gopinath, and many, many others.

The late 1990s and early 2000s also saw the rebirth of anti-assimilationist ideas. Michael Warner's 1999 book, *The Trouble with Normal,* questioned the gay rights movement's focus on marriage, an institution that would allow gay people to be more "normal," shaming those who did not want this kind of life. Mattilda Bernstein Sycamore made similar arguments in her 2004 book, *That's Revolting: Queer Strategies for Resisting Assimilation.* She pushed against mainstream gay and lesbian movements for continuing to fight for marriage rights rather than universal healthcare and housing for all.

An important concept within queer theory in the early 2000s was the idea of "pinkwashing." This phrase called out marketing and political strategies put in place to make a company or country appear more progressive and tolerant by emphasizing its support for gays and lesbians. A number of scholars have argued that Israel participates in pinkwashing through public demonstrations of its work for LGBTQ equality, while continuing to oppress its Palestinian minority. Women's and Gender Studies professor Jasbir Puar's 2007 book, *Terrorist Assemblages: Homonationalism in Queer Times*, makes the argument that the United States uses its supposed openness about homosexuality to disparage Muslim countries and as an excuse to intervene militarily, contending that it is standing up for women and LGBTQ people there.

What is Trans Studies?

In 2019, Ardel Haefele-Thomas and Thatcher Combs published the first book-length *Introduction to Transgender Studies*. However, trans studies has been a field much longer than it has been formally defined.

Like queer theory, trans studies emerged out of a recognized need for new voices within academic worlds. For many years, scholarly work was being written about trans people rather than by trans people, and even within LGBTQ communities, the ideas of trans academics were pushed aside by mainstream gays and lesbians.

Sandy Stone's 1992 essay "The *Empire* Strikes Back: A Posttranssexual Manifesto" is said to have kicked off the field of trans studies. Stone was writing in response to cisgender feminist Janice Raymond's 1979 book *The Transsexual Empire: The Making of the She-Male*, which argued that trans women "rape women's bodies by reducing the real female form to an artifact, appropriating this body for themselves" (p. 104). Raymond's book publicly attacked Stone for working as a sound engineer at the all-women's Olivia Records, accusing her of infiltrating women's spaces.

In her response, rather than attacking back, Stone adeptly brought to light a number of critical issues affecting trans lives. She pointed out that trans bodies were being discussed by medical professionals and feminist thinkers, who were "meeting on the battlefield of the transsexual body" (p. 230), without an understanding of trans people's lived experiences. The essay also explored pathologization of trans identities, gatekeeping by medical professionals, and the idea that trans bodies, rather than reinforcing the gender binary, have the "potential for productive disruption," that can lead us to more interesting and exciting ways of living. Most importantly, she argued that "a counterdiscourse is critical" (p. 230), thus officially launching the field of trans studies.

During the early period of Trans Studies, in the 1990s, it was slightly easier than it is today to keep track of emerging writers in the field. Leslie Feinberg (1997) introduced the world to gender nonconforming people in history in *Transgender Warriors: Making History from Joan of Arc to Dennis Rodman*, Jack Halberstam (1998) explored the ways that masculinity played out on traditionally female bodies in *Female Masculinity*, and Kate Bornstein (1997) helped readers to connect to their own gender identities in *My Gender Workbook: How to Become a Real Man, a Real Woman, the Real You, or Something Else Entirely*. Towards the end of the 1990s, trans theorists began to push boundaries even more, examining the progress of political movements (erotica author Patrick Califia, 1997, in *Sex Changes: The Politics of Transgenderism*) and the potential for more radical conceptualizations of identity (Riki Anne Wilchins, 1997, in *Read My Lips: Sexual Subversion and the End of Gender*).

The turn of the century brought a veritable boom in trans scholarship. It would be impossible to name even all of the major works published during this time. Legal scholars Paisley Currah, Richard Juang, and Shannon Minter (2006) created the first primer on trans legal issues (*Transgender Rights*), while Susan Stryker (2008) took on the broad subject of *Transgender History*. Scholar-activist Mattilda Bernstein Sycamore (2006) addressed trans attempts to fit into the mainstream in *Nobody Passes: Rejecting the Rules of Gender and Conformity*. Julia Serano (2007) explored the specific experiences of transfeminine people in *Whipping Girls: A Transsexual Woman on Sexism and the Scapegoating of Femininity*. At the cusp of the second decade of the 21st century, Kate Bornstein and S. Bear Bergman (2010) brought us *Gender Outlaws: The Next Generation*.

The 2006 publication of the *Transgender Studies Reader*, edited by Susan Stryker and Stephen Whittle, was a momentous accomplishment, bringing together a wide range of texts, beginning with excerpts from medical textbooks written by cisgender professionals in the early 1900s and then sampling from feminist, queer, and trans authors from a diverse array of backgrounds. It was one of the first major texts to explicitly address intersections between race, nationality, and gender.

The second decade of the 21st century was a turning point in trans scholarship. In 2014, *Transgender Studies Quarterly*, helmed by Susan Stryker and Paisley Currah and published by Duke University Press, debuted as the first nonmedical Transgender Studies journal, and, in 2015, the University of Arizona became the first ever institution to offer a masters in Transgender Studies. In 2014, a team of more than 50 authors published *Trans Bodies, Trans Selves* (Erickson-Schroth, 2014), a resource guide written by and for trans people, based on the well-known cisgender women's book *Our Bodies, Ourselves*.

In the second decade of the 21st century, Trans Studies has blossomed to include many new voices addressing issues that had been missing from academic discussions. For years, the field of trans studies benefitted from Black and Brown and indigenous cultures and ideas, but few of these contributions showed up in formal scholarship. The 2014 book *decolonizing trans/gender 101* (binaohan, 2014) attempts to move beyond 101 conversations and away from the centering of white trans narratives. C. Riley Snorton's (2017) *Black on Both Sides: A Racial History of Trans Identity* meticulously investigates the presence and meaning of Black gender nonconforming identities over time. In *Captive Genders: Trans Embodiment and the Prison Industrial Complex* (Stanley & Smith, 2015), essays address the surveillance and policing of gendered bodies, especially those of trans people of color. *Trap Door: Trans Cultural Production and the Politics of Visibility* (Gossett, Stanley, & Burton, 2017) wrestles with the contradiction that, as trans representation in media has increased dramatically, the everyday lives of trans people have, in many ways, not improved, and, in some ways, become increasingly fragile and prone to both state and individual violence.

Across the globe, trans identities are being explored in every culture, although this exploration does not always reach western-centric academic thought. Some attempts have been made to incorporate trans academic work from other areas of the world. For example, *Trans Studies Quarterly* published an issue on trans academic work in Latin America. However, the majority of what is considered Trans Studies remains white- and euro/U.S.-centric.

Trans Studies is a young field, but growing at an infinitely fast pace. Scholars of the next decade are likely to bring with them new and exciting takes on the possibilities for the future of gender.

What is transfeminism?

Transfeminism (also referred to as trans feminism) works to apply a gender-inclusive approach to feminist work and a feminist approach to trans discourse. Transfeminists, like other marginalized groups within feminism, such as working class women and women of color, have challenged the idea that there is a universal experience of being a woman.

Emi Koyama first published *The Transfeminist Manifesto* in 2001, but trans people have been involved in feminist work since it began. There are many important issues that unite cisgender women and trans people, including gender equality, violence prevention, and reproductive rights.

In the 1970s, trans women participated in a number of women's organizations. Some were warmly accepted by cisgender feminists, while others were rejected simply based on their sex assigned at birth. Beth Elliot, a transgender folk singer and activist, served as vice president and editor of the newsletter for the San Francisco chapter of the lesbian organization Daughters of Bilitis from 1971 to 1972. When she first joined, she was met with skepticism, but ultimately welcomed into the group. However, in late 1972, she was forced out in a vote that banned any trans women from participating.

In these early days, trans men, many of whom had spent significant portions of their lives in lesbian communities prior to coming out as trans, also often had strained relationships with feminist groups and were sometimes branded "traitors" for transitioning.

At the turn of the 21st century, trans people continued to run up against discriminatory practices in feminist spaces. A well-known example is the 1991 ejection of a trans woman named Nancy Burkholder from the Michigan Womyn's Music Festival, which maintained a women-born-women policy. The response by trans activists was to pitch tents nearby in what they called Camp Trans, an inclusive space for people of all gender identities, in protest of the ban. Clashes continued until the festival ultimately shuttered in 2015.

At certain events today, anti-trans sentiment continues, and trans attendees are treated with hostility. However, many feminist events are now open to people of all genders.

In formal academic circles, prior to the appearance of the term *transfeminism*, many trans authors already came to their work from a feminist standpoint. Certainly, turn-of-the-century trans authors such as Kate Bornstein, Paisley Currah, Leslie Feinberg and Jack Halberstam, approached the world through a feminist lens. Max Wolf Valerio contributed to both *This Bridge Called My Back* (Moraga & Anzaldúa, 1981) and *This Bridge We Call Home* (Anzaldúa & Keating, 2013), anthologies addressing feminism and people of color, the first prior to his transition and the second, after.

Diana Courvant, a trans activist, may have coined the term *transfeminism* in the early 1990s. Patrick Califia is thought to have been the first to refer to transfeminism in a full-length book when he published *Sex Changes: The Politics of Transgenderism* in 1997. Jessica Xavier also used the term in her 1999 article "Passing as Privilege." However, it was Emi Koyama's 2001 "Transfeminist Manifesto" that put the term on the map.

Koyama's essay defines transfeminism as "a movement by and for trans women who view their liberation to be intrinsically linked to the liberation of all women" (p. 245). She takes a straightforward approach to issues facing trans women, beginning with the difficulties they run into attempting to be a part of feminist groups. She points out that, like other marginalized women, trans women have been accused of fragmenting feminism, but instead expand the boundaries of feminist work. Koyama also asserts that differences between cisgender and transgender women should be acknowledged, including the ways that trans women have benefited from male privilege and cis women have benefitted from cis privilege. She stresses the importance of working together to address important issues such as violence against women and reproductive choice.

Since the publication of the "Transfeminist Manifesto," a number of other authors have produced work related to transfeminism. One of the most well-known individual writers to address feminism from a trans standpoint is Julia Serano, whose 2007 *Whipping Girl: A Transsexual Woman on Sexism and the Scapegoating of Femininity* argues that gender policing directed at trans women is a unique form of misogyny that Serano calls transmisogyny.

Anthologies have included Krista Scott-Dixon's 2006 *Trans/Forming Feminisms: Transfeminist Voices Speak Out* as well as Anne Enke's 2012 *Transfeminist Perspectives in and Beyond Transgender and Gender Studies*. The May 2016 issue of *Transgender Studies Quarterly* was titled *Trans/Feminisms*. The first volume of the *Transgender Studies Reader* (2006) included Emi Koyama's essay "Whose Feminism Is It Anyway? The Unspoken Racism of the Trans Inclusion Debate," and the second volume (2013) expanded on this to incorporate an entire section on transfeminism.

In 2019, author Andrea Long Chu published *Females*, a crossover memoir/theoretical treatise in which, instead of arguing that trans women should be considered women, she contends that we are *all* women. In Chu's view, the term *female* has come to mean "any psychic operation in which the self is sacrificed to make room for the desires of another" (p. 11). "Everyone is female," she states, "and everyone hates it" (p. 11). Chu's name rose in recognition with the publication of her November 2018 *New York Times* Op-Ed, "My New Vagina Won't Make Me Happy." In it, she approaches the medical model of transition with skepticism, revealing that she believes hormones and surgeries make a difference in trans people's lives but that outcomes should not be solely based on happiness or improvement in mental health symptoms because these measures do not necessarily improve with physical transition and are often used by medical professionals to deny care to those who they do not feel are likely to show these improvements.

Other transfeminists have disagreed with Chu's take. One is Kai Cheng Thom, a Canadian social worker who has published both fiction and nonfiction books. In her response piece in *Slate* magazine, titled "The Pain—and Joy—of Transition," Thom agrees with Chu that the right to transition should not be based on presumed mental health outcomes but instead on bodily autonomy and informed consent. However, she questions Chu's belief that there are "no good outcomes in transition," pointing out that multiple large studies show improvement in mental health after transition. Thom also makes a plea for trans communities to come together to support each other even when their approaches differ.

Because many trans authors today fall into the category "feminist" but do not use the term *transfeminist*, there are likely many more transfeminist voices today than a simple Google search will reveal. Hopefully this means that there is much more exciting transfeminist scholarship to come.

7

THE FUTURE OF GENDER

How might reproductive technologies revolutionize parenting?

If scientists from just a few decades ago were to visit a fertility clinic today, they would be astonished by the progress that has been made. The first baby created through in vitro fertilization, the process by which an egg and sperm are combined outside of the body, was born in 1978. Since then, new developments have allowed for genetic testing of embryos, surrogate carriers, three-parent genetic babies, and even uterine transplants. In 2017, a woman gave birth to a baby conceived from an embryo created 24 years prior—only one year after the mother herself had been born.

The prospects for future reproductive technologies are almost beyond our imaginations. Science fiction movies have suggested for years that artificial wombs will one day take over as our main means of reproduction. Transgender women and cisgender men may one day be able to have uterus transplants and give birth. While we are not quite there yet, it does seem likely that before too long we will have that capability, as there have now been successful uterine transplants in cisgender women.

So what does the future of reproductive technology look like for the next few decades? One thing that is fairly certain is that we are going to have much improved genetic testing of embryos created through in vitro fertilization. Preimplantation genetic screening is already being used to test for chromosomal disorders such as Down syndrome, where there is a third copy of chromosome 21, and to identify the sex of the embryo. Preimplantation genetic diagnosis can identify mutations that cause certain genetic diseases.

The 1997 film *Gattaca* depicted a world in which the majority of people were created through genetic selection of embryos. The main character, Vincent, was born through "natural" means and, like other genetically inferior "invalids," is relegated to menial tasks and expected to have a short lifespan. His adventure begins when he uses DNA samples from a "valid" to pose as a different breed of citizen.

The fictional future of *Gattaca* seems a stretch, but it is likely that within just a decade or two, genetic testing will be able to predict many more characteristics of an embryo—eye and hair color, height, weight, and certain risk factors, such as having a propensity for developing diabetes or heart disease. Perhaps just a decade or two later, we may be able to not just screen embryos, but to alter their genetic material to insert desired characteristics. In addition, as these screening techniques become better understood, they will likely also become more affordable, making them available to a larger portion of the population, and, at a certain point, expected. As in *Gattaca*, it is conceivable that one day we could live in a world where those who have not been created through genetic selection live as second-class citizens.

Another area of reproductive technology that is likely to significantly change our understanding of parenting is the creation of eggs and sperm from stem cells. Currently, we are limited to the eggs and sperm our own bodies can produce or to using donated eggs or sperm. However, scientists have now been able to take stem cells from mice and push them to differentiate into sperm cells, then inject these into mice eggs to create embryos that grow into healthy mice.

The creation of "artificial" sperm and eggs could open up possibilities for all kinds of changes in our reproductive capacity. On a basic level, it could allow those who are infertile, either because of low sperm count or age-related changes in eggs, to create brand-new viable sperm or eggs. Men with nonfunctioning sperm or women in their 50s or 60s could be capable of having biological children. But even beyond that, stem cells from someone of any sex could theoretically be used to create either sperm or eggs, allowing transgender men to produce sperm or transgender women to produce eggs if they wanted. It could also allow same-sex couples to have biological children together.

What is gender-neutral parenting?

> Once upon a time, a baby named X was born. This baby was named X so that nobody could tell whether it was a boy or a girl. Its parents could tell, of course, but they couldn't tell anybody else. They couldn't even tell Baby X at first. You see, it was all part of a very important Secret Scientific Xperiment, known officially as Project Baby X. (Gould, 1978, p. 1)

So begins Lois Gould's 1978 children's book *X: A Fabulous Child's Story*. In the book, a couple, the Joneses, work with scientists to raise a baby outside the confines of gender stereotypes, not telling friends or neighbors the baby's sex. At first, there is significant pushback. Friends and neighbors don't know what to buy for the baby. Store clerks tell the Joneses they can't help them pick out clothing or toys. The Joneses worry about sending baby X to school. But when the other children see how liberating it is for X to be able to play with dolls, but also be star quarterback and to win relay races alongside baking contests, they all decide they want to be just like X. Despite the children's excitement, the other parents remain wary, but eventually give in and welcome the Joneses to the parents' association.

Written at the tail end of second wave feminism, Gould's book represents an idealistic approach to parenting with the goal of providing as many opportunities as possible for a child while avoiding gender stereotypes. In the early 1970s, prior to its publication, numerous studies had been conducted that stressed the ways in which parents and other adults treated children differently based on gender, leading to disparities in children's intellectual, physical, and emotional abilities. Girls lagged behind in science, math, and sports, while boys lacked verbal and emotional skills.

Gender stereotypes and their consequences are not a thing of the past. Sociologist Elizabeth Sweet has analyzed over 7,000 toy advertisements in Sears catalogues and found that toys are more gendered today than at any other point in the past century. The 2017 Global Early Adolescent Study found that girls who conform to gender stereotypes are more likely to be depressed, leave school early, and be exposed to violence. Boys who conform to gender stereotypes have a shorter life expectancy, are more likely to engage in physical violence, and are more prone to substance abuse and suicide.

Given the amount of harm caused by gender stereotypes, many parents today attempt to shield their children from the negative effects and encourage them to explore a variety of activities and interests. However, even with conscientious parents, outside influences can be substantial. Because of this, some parents have taken a stance similar to the Joneses, raising their children gender neutral or genderless, not revealing the sex of the child until later in life.

Genderless or gender-neutral parenting can include choosing a gender neutral name and using gender neutral pronouns like *they*, instead of *he* or *she*. Some families may buy clothing and toys that are gender-neutral. Others may alternate between stereotypically masculine and feminine items or dress the child in multiple items of clothing, some of which are considered more masculine and some more feminine. There are Facebook groups to connect parents who are raising their children gender-neutral. Some parents have social networking accounts to show others what they have learned. Working with daycare programs and schools can be challenging. School activities as simple as lining up or choosing seating arrangements may be gendered. Parents often talk to their child's school in advance to ensure that as much as possible can be done so that gender stereotypes are not perpetuated.

There has been a significant amount of controversy about the idea of gender-neutral parenting and derision directed toward parents who decide to take this route, especially when the decision is made public outside of a neighborhood or small community. Although children, particularly when they are young, can be very accepting of new ideas, adults can find it harder. When the Canadian couple David Stocker and Kathy Witterick announced that they would be raising their child, Storm, without revealing Storm's sex, people delivered angry letters to their door and shouted at them through car windows.

Many parents raising their children gender-neutral feel that the most difficult part is managing other people's reactions. Close relatives often feel left out if they are not told the child's sex. Friends, neighbors, and even total strangers express concerns about how the child will develop. One common anxiety is that the child is being forced into a lifestyle that the child did not choose. Parents who elect gender-neutral parenting argue that they are doing just the opposite—providing their children with more rather than fewer opportunities, allowing them the freedom to explore and choose what is best for them, without social pressure. Critics also state that children raised this way will be confused about their sex, gender, or both. Parents who use gender-neutral parenting respond by emphasizing how important it is to have conversations with children about these concepts.

Despite the media attention, true gender-neutral parenting is still rare. However, many parents attempt to be as gender-neutral in their parenting as possible, even if they assign their child a gender at birth. They may encourage their child to try clothing and toys traditionally associated with different genders and allow their child to pick out their own clothing for school, even if it does not match gender expectations. They may talk with their child about sex and gender at an early age and stress the history of sexism and rigid gender stereotypes that lead to harm for people of all genders.

In some parts of the world, these approaches are broadening outside of individual families to enter into day care programs and schools. In Sweden, preschool teachers avoid the term "boys and girls," calling children "friends." A gender-neutral Swedish pronoun, *hen*, was introduced in 2012 and allows teachers to use one word for all children, instead of either *he* or *she*. These techniques seem to be making a difference. Children in schools that use gender-neutral language are less likely to make assumptions based on gender stereotypes and more likely to have friends of another gender. Whether or not a particular family chooses to use gender-neutral parenting, approaches to child-rearing that intentionally break down stereotypes can benefit all children.

How might new technologies shape our ability to modify the human body?

Technology has always aided the desire of humans to amplify gender-conforming aspects of the body, be it through the early use of make-up by the ancient Egyptians, the popularization of the corset throughout the 16th century, or access to breast augmentation in the 20th century. Compared to only a few decades ago, there has been significant medical and social progress in the arena of gendered body modification, particularly for transgender communities. Hormones have become more accessible and, in the United States, can often be obtained through informed consent models that allow people, as long as they are judged fit to make medical decisions for themselves, to start hormones when they feel they are ready.

In the past, most clinics that provided hormone therapies had strict criteria for prescribing medication, including that those seeking help provide stereotypical accounts of their childhoods and identify as straight after transition. This meant that far fewer people were able to medically transition, and many who did felt that they had to lie at least about some aspects of their experiences.

Since the beginning of the 21st century, one of the biggest breakthroughs in hormone treatment has been the use of puberty blockers to delay puberty in transgender youth. Puberty blockers have allowed some transgender young people to avoid experiencing an unwanted puberty and have eliminated their need for later body modifications they could require to reverse that puberty. Studies report few negative side effects from these medications.

Interestingly, it was not a medical discovery that led to this breakthrough, as puberty blockers had existed since the 1980s and were readily employed to treat children with conditions that activated early puberty. Instead, it was an increase in social recognition and acceptance of transgender youth that led to more physicians being trained in and willing to use puberty blockers, as well as a select group of insurance plans changing their policies to cover these medications.

While hormone treatment can have desired gendered effects on the body, such as changes in body fat and muscle distribution and body/facial hair, there are body parts for which surgery is the most effective method of change. Improvements in surgery techniques have led to results that are more aesthetically pleasing and preserve functioning, such as nipple and genital sensation. Like other medical treatments for transgender people, an increasing number of U.S. insurance companies are now also covering certain types of gender-affirming surgeries.

Breast augmentation for transgender women and chest reconstruction—or top surgery—for transgender men, are generally extremely effective surgeries. Facial feminization surgeries have advanced to the point where physicians are able to reverse many of the facial structure changes that occur with male puberty. Body contouring surgeries can enhance the fat located in certain areas of the body or remove it, helping achieve more or less curvy silhouettes that are more gender-typical based on cisnormative body ideals.

Genital surgeries for transgender individuals include a variety of approaches and options and continue to modernize as more complex techniques are developed to enhance circulation, sensation, and appearance. Vaginoplasty, the creation of a vagina, is performed by different surgeons using varied techniques and is generally successful aesthetically as well as in preserving urinary function and genital sensation. Phalloplasty, or the creation of a penis, is considered a more complex procedure as it requires tissue from another area of the person's body to create the phallus and specific techniques to lengthen the urethra outside the body. While some trans men, as well as cisgender men whose genitals have been injured, do undergo full phalloplasties, there are many alternative methods available for men interested in genital surgery. Some transmasculine individuals opt for a metoidioplasty, which cuts the clitoral ligament so that the clitoris pushes farther outside of the body as a phallus. A special kind of metoidioplasty called a ring metoidioplasty extends the urethra through the elongated clitoris, allowing the person to stand and pee. Some people who undergo phalloplasties and metoidioplasties also have a vaginectomy, which closes the vaginal canal, while others do not.

In 2014, surgeons in South Africa performed the world's first successful penis transplant, paving the way for the possibility of a new type of genital construction. Between 2014 and 2018, there were three other successful transplants, all on cisgender men, including a U.S. soldier whose genitals had been destroyed in combat.

Looking ahead, there is a world of possibilities for how new technologies could allow us to improve on existing gender-related body modifications. Surgical techniques, especially, are likely to change dramatically over the next decade or two. We might progress from using our own or donor tissue for transplantation to substituting more sophisticated synthetic materials, or growing the desired tissues from stem cells. Prosthetic hips and knees may make way for prosthetic genitals. Creating transferrable body parts from our own DNA may be as easy as 3-D printing an organ that can be attached to our existing nerves and function like any of our other organs. Will these technologies end up making us less human? Or will they allow us to be even more of ourselves?

What are the current frontiers in gender equality?

When we talk about frontiers, we generally have expansive ideas in mind. In terms of gender equality, we may picture women in positions of world power, such as presidents or prime ministers, or women as the pinnacles of success, making scientific discoveries and writing the next brilliant work of fiction. Women are pushing these boundaries every day.

Seventy of 146 nations studied by the World Economic Forum have now had a female leader. In recent years, almost 50% of the books on the New York Times Bestsellers list have been written by women. Nearly half of tenure track professors in the United States are women. In 2018, New Zealand passed legislation granting victims of domestic violence ten days paid leave, allowing them to leave their partners, protect themselves and their children, and seek out new housing. In January 2019, 5 million women in India formed a 385-mile human protest for gender equality. Across the globe, women, girls, and transgender people are organizing and advocating for change.

At the same time, women continue to live in starkly different economic conditions than men. Around the world, women are more likely than men to live in poverty. Sixty percent of chronically hungry people are women and girls. In the United States, women are only 5% of Fortune 500 CEOs, but 63% of those earning minimum wage. Globally, women are paid 23% less than men. Only 20% of the world's landowners are women. Seventy-five percent of women in developing countries work in the informal economy, without access to legal rights or protections. Women do twice as much unpaid work, such as household chores and childcare. The United Nations Foundation estimates that unpaid work done by women across the globe, if paid, would add up to $10 trillion a year.

Despite many important changes in the status of women over the last two centuries, women and girls continue to live in some of the most miserable conditions in the world. Images of women in positions of power in governments and academic settings is inspiring, and the appearance of women astronauts and physicists makes for exciting news. However, many experts and aid organizations suggest that there is an urgent need to direct resources and attention to creating a more equitable society. With regards to economic parity, it is estimated that the gender gap will take 257 years to close. Today, there are 72 countries where women are barred from opening bank accounts or obtaining credit, and there is no country where men spend the same amount of time on unpaid work as women.

Two thirds of illiterate people in the world are women. Every day, millions of girls, especially in rural areas, watch as their brothers leave for school. This is despite the fact that we know that educating girls works. Every year of primary school increases a woman's eventual wages by 10% to 20%. Women who have been educated have lower rates of early pregnancy, death in childbirth, and child malnutrition.

Feminists have argued for years that women having control over when and how to have children is essential to gender equality. Some have even suggested that until reproduction is separated from human bodies (e.g., through the use of artificial wombs), women will never have the same opportunities as men. While artificial wombs may be a long way off and may not be desirable for everyone, granting women the ability to control their own reproduction may have similar results. Studies show that access to contraception allows women to further their education and move forward in their careers. Access to safe abortions similarly allows for increased gender equity. However, in 2018, over 40% of women lived in countries with highly restrictive abortion laws.

Time out of work or school to have children greatly affects women's economic potential. Once someone has decided to become pregnant, in many areas of the world, there is no such thing as maternity leave, and in those with this benefit, it is often unpaid. Opponents of comprehensive maternity leave argue that women choose to become pregnant and therefore should shoulder the burden themselves. This argument falls flat when we consider that no children would ever be born if no one gave birth to them.

In addition to the effects of maternity leave itself, women also take on the lion's share of childcare and household work, often leaving them with time for part-time work only, or working long hours at paid work and then returning home to work even longer at home. Women who stay home with children while male partners work have little access to economic freedoms that would allow them to make their own choices about their lives. Some countries, particularly those with democratic socialist governments, have introduced reasonable lengths of paid maternity and paternity leave, as well as universal childcare so that women have improved access to education and career advancement.

A frontier that is less often considered, but could be key to reaching gender equity is the promotion of typically unpaid women's work (including child and elder care, cooking, cleaning, and running errands) as a shared societal burden that is either spread out over both genders or monetarily compensated. Women perform about two and half times more of this kind of work globally than men do, and even in countries where social norms encourage men to be more involved, such as Sweden, women still perform a third more of the unpaid work. Making this type of work more equitable or receiving pay for it could drastically change household economics.

Healthcare access is stratified not only by geographic location and class, but also gender. The United States has the worst maternal mortality rate of any developed country. Despite global maternal death rates falling by more than one third from 2000 to 2015, outcomes for American mothers worsened. African American, Native American, and Alaskan Native women die of pregnancy-related causes at a rate about three times higher than those of white women. For transgender individuals, access to basic healthcare and gender-affirming interventions is often a laborious process marked by roadblocks, expense, and discrimination. The United States 2015 Transgender Health Survey, the largest of its kind, found that the combination of anti-transgender bias and persistent, structural racism was especially devastating to the health and wellness of transgender individuals. The survey found that transgender people faced higher rates of HIV infection, smoking, drug and alcohol use, and suicide attempts than the general population. For transgender and gender diverse individuals, limited access to safe housing and education, precarious legal protections, violence, and discrimination continue to threaten lives and livelihoods globally.

Although much progress has been made toward gender equality across the globe, major power disparities still exist today. Thinking optimistically about the future, we can imagine a world where gender inequality is a thing of the past—where individuals of all genders have equal rights, access to the same opportunities, and are treated with the same respect and autonomy. However, realistically, the work that remains would require monumental restructuring of institutions, the undoing of strongly entrenched biases, and the rethinking of long-standing cultural assumptions.

Could the future be genderless? Do we want it to be?

What would a genderless future look like? Some people might imagine a world where humans no longer have bodies that signify gender—one in which we are all androgynous. Medical technologies may one day advance to the point where this is possible. But is it probable?

With the myriad of varied gender identities and presentations that exist today, it seems unlikely that we would all soon conform to a standard androgynous body and more likely that definitions of gender and experiences of gendered bodies would instead expand. But what would it mean to expand our definitions of gender? How exactly would that work?

While we almost certainly cannot know the answers to these questions now, we can imagine what the future of gender might look like in different scenarios.

In the future, it is possible (although not probable) that anti-oppressive social forces would be so successful in fighting male privilege that masculinity would no longer be advantaged over femininity, men and women would have equal earnings, and gender inequality would cease to divide and disenfranchise entire swaths of the population. Our bodies would be the same as they are now, but birth certificates would not have genders because they would be useless. Children would be raised without particular expectations for boys or girls and would be supported in whatever styles of dress or activities they preferred. As adults, all people would have access to all kinds of careers and partners, regardless of gender.

Despite the many gains we have made, this scenario seems unlikely given the weight of ingrained social attitudes about gender. But what if technology changed the way we interacted?

In another future scenario, we could all become so integrated into cyber technology that we would live most of our lives online. Virtual worlds like Second Life that allowed people to create avatars and interact with each other through those avatars would become our primary modes of communication. In this kind of future, it is possible that we could all choose any kind of avatar we wanted, without concern for how our gendered bodies interfacing with our computers were actually configured. Our avatar bodies could be like ours, different from ours, transgender, cisgender, or intersex. In these virtual worlds, we could be whatever we wanted to be. What would it mean to communicate, work, and even develop relationships with others with identified genders untethered from our flesh and blood bodies behind the screen? Would that alter the way we think about our own identities? Could it change the amount of importance we place on gender and the assumptions we make based on gender?

Although we have access to virtual worlds now, most of us continue to spend the majority of our time interacting with others in the "real" world, so it would likely take a significant change in social norms or a consequential upgrade in virtual technology to bring us to the point where we interacted almost completely through virtual worlds. But what if the technology change involved our bodies and not our modes of communication?

Every day we move closer and closer to the ability to replace our body parts with either artificial parts made of metal or plastic or with human body parts grown from stem cells. At a certain point, it may become so normal to replace body parts that we may all essentially be cyborgs—part human and part mechanical. Technology may lead us to be almost immortal. Although it might be difficult to prevent accidents, when our organs began to fail, we would simply replace them. What would this mean for our gendered selves? One thing it might mean is that we would have the freedom to try out differently-gendered bodies. And if anyone could change the gender of their body, what would gender mean? Or would it become meaningless?

While the language, norms, and expectations around gender have shifted over the past century to expand our understanding of the vast nature of gender and gains in technology provide the potential for an even more limitless, variable experience of gender, the future of gender may not be as liberated and self-determined as some may hope. In 1985, Margaret Atwood published *The Handmaid's Tale,* and, in 2017, Hulu released a television series based on the novel. The plot presents a dystopian future wherein stark lines around gender are reinforced by biology. Particular to this story is a focus on reproductive rights and the governing powers that regulate women's bodies.

The novel and television show are fictional, but the depictions of gender-based violence, revocation of personal freedoms, and lack of access to education are common realities faced by women, girls, and other gender minorities across the world today. In 2017, Russia's parliament voted 380–3 to decriminalize domestic violence in cases where it did not cause "substantial bodily harm" and did not occur more than once a year. In 2018, Tokyo Medical University marked down the test scores of young women applying to medical school, limiting women to just one third of the class to ensure more men than women became doctors. The reason given was that authorities were concerned with women's ability to continue to work after starting a family, reinforcing a limit to women's earning power and ability to self-determine their careers. In May 2019, Alabama's state senate passed a near-total ban on abortion, making it a crime to perform the procedure at any stage of pregnancy, with no exception for rape or incest. Internationally, the global gag rule prohibits foreign nongovernmental organizations that receive funding from the U.S. government from providing abortion services or referrals, while also barring advocacy for abortion law reform, even if it's done with the nongovernmental organization's own, non-U.S. funds. These realities make it difficult to imagine a genderless world, particularly because gender and power are so dynamically related in our current societies.

It is possible that if we continue to make progress, we may be able to move toward freeing ourselves from the confines of gender oppression and gender-based violence, into a future where gender becomes less of a determinant of safety, success, and freedom. Technological advancements may render bodies more gender fluid than we have known to be possible, and body-based gender versus our online presence may reduce or eliminate the gendered focal point of identity. Advancements in reproductive technologies may change the biologically deterministic ideology of our global economy, health, education, and culture. Yet there exist many concerning trends today that call into question our ability to make the type of progress we would hope for. True, we could be moving toward the possibility of a genderless future, where more freedom exists to explore diverse gender experiences, aided by technology and medicine. However, this experience, at least at first, will likely be reserved for the most privileged. Technological advancements that could be shifting us toward a potential genderless future may not parallel social movements toward equality. A true genderless future may therefore only be an option once gender equality is achieved, necessitating ongoing dedication to education and advocacy.

BIBLIOGRAPHY

Chapter 1

Baba, T., Endo, T., Honnma, H., Kitajima, Y., Hayashi, T., Ikeda, H., . . . Saito, T. (2007). Association between polycystic ovary syndrome and female-to-male transsexuality. *Human Reproduction, 22*(4), 1011–1016.

Bianchi, S. M., Sayer, L. C., Milkie, M. A., & Robinson, J. P. (2012). Housework: Who did, does or will do it, and how much does it matter? *Social Forces, 91*(1), 55–63.

Blackless, M., Charuvastra, A., Derryck, A., Fausto-Sterling, A., Lauzanne, K., & Lee, E. (2000). How sexually dimorphic are we? Review and synthesis. *American Journal of Human Biology, 12*(2), 151–166.

Brydum, S. (2015, November 6). Drop the T. *Advocate.* Retrieved from https://www.advocate.com/transgender/2015/11/06/lgbt-groups-respond-petition-asking-drop-t

Donner, F. (2020, February 12). The household work men and women do, and why. *New York Times.* Retrieved from https://www.nytimes.com/2020/02/12/us/the-household-work-men-and-women-do-and-why.html

Erickson-Schroth, L. (Ed.). (2014). *Trans bodies, trans selves: A resource for the transgender community*. New York, NY: Oxford University Press.

Hill, M. R., Mays, J., & Mack, R. (2013). *The gender book*. Houston, TX: Marshall House Press.

Mancini, Talia. Ms. (2017, October 11). A new study shows how gender stereotypes hurt kids around the world.

Retrieved from http://msmagazine.com/blog/2017/10/11/new-study-shows-gender-stereotypes-hurt-kids-around-world/

Mmari, K., Blum, R. W., Atnafou, R., Chilet, E., De Meyer, S., El-Gibaly, O., . . . Zuo, X. (2017). Exploration of gender norms and socialization among early adolescents: The use of qualitative methods for the Global Early Adolescent Study. *Journal of Adolescent Health, 61*(4), S12–S18.

Mueller, A., Gooren, L. J., Naton-Schotz, S., Cupisti, S., Beckmann, M. W., & Dittrich, R. (2008). Prevalence of polycystic ovary syndrome and hyperandrogenemia in female-to-male transsexuals. *Journal of Clinical Endocrinology & Metabolism, 93*(4), 1408–1411.

Paoletti, J. B. (2012). *Pink and blue: Telling the boys from the girls in America*. Bloomington, IN: Indiana University Press.

Pew Research Center. (2015, November 4). Raising kids and running a household: How working parents share the load. Retrieved from https://www.pewsocialtrends.org/2015/11/04/raising-kids-and-running-a-household-how-working-parents-share-the-load/

Teich, N. M. (2012). *Transgender 101: A simple guide to a complex issue*. New York, NY: Columbia University Press.

World Economic Forum. (2017). *The global gender gap report*. Geneva, Switzerland: Author.

Chapter 2

Devlin, H. (2015, May 14). Early men and women were equal, say scientists. *The Guardian*. Retrieved from https://www.theguardian.com/science/2015/may/14/early-men-women-equal-scientists

Dong, Y., Morgan, C., Chinenov, Y., Zhou, L., Fan, W., Ma, X., & Pechenkina, K. (2017). Shifting diets and the rise of male-biased inequality on the Central Plains of China during Eastern Zhou. *Proceedings of the National Academy of Sciences, 114*(5), 932–937.

Dyble, M., Salali, G. D., Chaudhary, N., Page, A., Smith, D., Thompson, J., . . . Migliano, A. B. (2015). Sex equality can explain the unique social structure of hunter-gatherer bands. *Science, 348*(6236), 796–798.

Feinberg, L. (1996). *Transgender warriors: Making history from Joan of Arc to Dennis Rodman*. Boston, MA: Beacon Press.

Gettleman, J. (2018, February 17). The peculiar position of India's third gender. *New York Times*. Retrieved from https://www.nytimes.com/2018/02/17/style/india-third-gender-hijras-transgender.html

Goettner-Abendroth, H. (2012). *Matriarchal societies: Studies on indigenous cultures across the globe*. New York, NY: Peter Lang.

Gossett, R., Stanley, E. A., & Burton, J. (2017). *Trap door: Trans cultural production and the politics of visibility*. Cambridge, MA: MIT Press.

Martin, W. (2018). *Untrue: Why nearly everything we believe about women, lust, and infidelity is wrong and how the new science can set us free*. New York, NY: Little, Brown Spark.

Nagle, R. (2018, June 30). Queer voices: The healing history of Two-Spirit, a term that gives LGBTQ natives a voice. *HuffPost*. Retrieved from https://www.huffpost.com/entry/two-spirit-identity_n_5b37cfbce4b007aa2f809af1

Rogers, D. (2012, July 25). Inequality: Why egalitarian societies died out. *New Scientist*. Retrieved from https://www.newscientist.com/article/dn22071-inequality-why-egalitarian-societies-died-out/

Snow, D. R. (2013). Sexual dimorphism in European Upper Paleolithic cave art. *American Antiquity*, *78*(4), 746–761.

Stryker, S. (2017). *Transgender history: The roots of today's revolution*. Berkeley, CA: Seal Press.

Synowiec, O. (2018, November 26). The third gender of southern Mexico. *BBC*. Retrieved from http://www.bbc.com/travel/story/20181125-the-third-gender-of-southern-mexico

Chapter 3

Brink, L. (2013, May 15). Why people keep misunderstanding the "connection" between race and IQ. *The Atlantic*. Retrieved from https://www.theatlantic.com/national/archive/2013/05/why-people-keep-misunderstanding-the-connection-between-race-and-iq/275876/

Brown, C. S. (2017, July 24). Everything you believe is wrong: There is no such thing as a male or female brain. *Fast Company*. Retrieved from https://www.fastcompany.com/40441920/everything-you-believe-is-wrong-there-is-no-such-thing-as-a-male-or-female-brain

Desjardins, B. (2004, August 30). Why is life expectancy longer for women than it is for men? *Scientific American*. Retrieved from https://www.scientificamerican.com/article/why-is-life-expectancy-lo/

Gebelhoff, R. (2017, January 26). We've been misled on the difference between genders. *Washington Post*. Retrieved from https://www.washingtonpost.com/news/in-theory/wp/2017/01/26/weve-been-misled-about-the-difference-between-genders/

Jordan-Young, R. M. (2011). *Brain storm: The flaws in the science of sex differences*. Cambridge, MA: Harvard University Press.
Lemaître, J. F., Ronget, V., Tidière, M., Allainé, D., Berger, V., Cohas, A., . . . Marais, G. A. (2020). Sex differences in adult lifespan and aging rates of mortality across wild mammals. *Proceedings of the National Academy of Sciences, 117*(15), 8546–8553.
Phoenix, C. H., Goy, R. W., Gerall, A. A., & Young W. C. (1959). Organizing action of prenatally administered testosterone propionate on the tissues mediating mating behavior in the female guinea pig. *Endocrinology, 65*, 369–382.
Robson, D. (2015, October 1). Why do women live longer than men? *BBC*. Retrieved from http://www.bbc.com/future/story/20151001-why-women-live-longer-than-men
Vigen, R., O'Donnell, C. I., Barón, A. E., Grunwald, G. K., Maddox, T. M., Bradley, S. M., . . . Rumsfeld, J. S. (2013). Association of testosterone therapy with mortality, myocardial infarction, and stroke in men with low testosterone levels. *JAMA, 310*(17), 1829–1836.

Chapter 4

Ainsworth, T. A., & Spiegel, J. H. (2010). Quality of life of individuals with and without facial feminization surgery or gender reassignment surgery. *Quality of Life Research, 19*(7), 1019–1024.
Association of Gay and Lesbian Psychiatrists. (n.d.). LGBT mental health syllabus. Retrieved from http://aglp.org/gap
Breslow, A. S., Brewster, M. E., Velez, B. L., Wong, S., Geiger, E., & Soderstrom, B. (2015). Resilience and collective action: Exploring buffers against minority stress for transgender individuals. *Psychology of Sexual Orientation and Gender Diversity, 2*(3), 253.
Drescher, J. (2015). Queer diagnoses revisited: The past and future of homosexuality and gender diagnoses in DSM and ICD. *International Review of Psychiatry, 27*(5), 386–395.
Gilman, S. L., King, H., Porter, R., Rousseau, G. S., & Showalter, E. (1993). *Hysteria beyond Freud.* Berkeley, CA: University of California Press.
Gonzales, G., & Henning-Smith, C. (2017), Barriers to care among transgender and gender nonconforming adults. *The Milbank Quarterly, 95*, 726–748.
Grady, D. (2017, December 2). Woman with transplanted uterus gives birth, the first in the U.S. *New York Times.* Retrieved from https://

www.nytimes.com/2017/12/02/health/uterus-transplant-baby.html

Kattari, S. K., Walls, N. E., Speer, S. R., & Kattari, L. (2016). Exploring the relationship between transgender-inclusive providers and mental health outcomes among transgender/gender variant people. *Social Work in Health Care, 55*(8), 635–650.

Khan, F. N. (2016, November 2016). A history of transgender health care. Retrieved from https://blogs.scientificamerican.com/guest-blog/a-history-of-transgender-health-care/

Klein, A., & Golub, S. A. (2016). Family rejection as a predictor of suicide attempts and substance misuse among transgender and gender nonconforming adults. *LGBT Health, 3*(3), 193–199.

Lazarus, S., & Folkman, S. (1984). *Stress, Appraisal, and Coping*. New York, NY: Springer.

Light, A. D., Obedin-Maliver, J., Sevelius, J. M., & Kerns, J. L. (2014). Transgender men who experienced pregnancy after female-to-male gender transitioning. *Obstetrics & Gynecology, 124*(6), 1120–1127.

Lindsay, K. (2016, August 24). How do trans men breastfeed their babies? *Cosmopolitan*. Retrieved from https://www.cosmopolitan.com/sex-love/a63217/chestfeeding/

Maines, R. (1999). *The technology of orgasm: "Hysteria," the vibrator, and women's sexual satisfaction.* Baltimore, MD: The Johns Hopkins University Press.

Mallory, C., Brown, T. N. T., & Conron, K. J. (2018, January). Conversion therapy and LGBT youth. *The Williams Institute*. Retrieved from https://williamsinstitute.law.ucla.edu/wp-content/uploads/Conversion-Therapy-LGBT-Youth-Jan-2018.pdf?response_type=embed

Malpas, J. (2011). Between pink and blue: A multi-dimensional family approach to gender nonconforming children and their families. *Family Process, 50*(4), 453–470.

Martin, C. L., Ruble, D. N., & Szkrybalo, J. (2002). Cognitive theories of early gender development. *Psychological Bulletin, 128*(6), 903–933.

Meyer, I. H. (1995). Minority stress and mental health in gay men. *Journal of Health and Social Behavior*, 38–56.

Meyer, I. H. (2003). Prejudice, social stress, and mental health in lesbian, gay, and bisexual populations: conceptual issues and research evidence. *Psychological Bulletin, 129*(5), 674.

Obedin-Maliver, J., Goldsmith, E. S., Stewart, L., White, W., Tran, E., Brenman, S., . . . Lunn, M. R. (2011). Lesbian, gay, bisexual, and transgender–related content in undergraduate medical education. *JAMA, 306*(9), 971–977.

Obedin-Maliver, J., & Makadon, H. J. (2016). Transgender men and pregnancy. *Obstetric Medicine, 9*(1), 4–8.

Olson, K. R., Durwood, L., DeMeules, M., & McLaughlin, K. A. (2016). Mental health of transgender children who are supported in their identities. *Pediatrics, 137*(3), 1–8.

Singh, A. E. (2017). *Affirmative counseling and psychological practice with transgender and gender nonconforming clients*. Washington, DC: American Psychological Association.

White Hughto, J. M., & Reisner, S. L. (2016). A systematic review of the effects of hormone therapy on psychological functioning and quality of life in transgender individuals. *Transgender Health, 1*(1), 21–31.

Whittle, S. (2010, June 2). A brief history of transgender issues. *The Guardian*. Retrieved from https://www.theguardian.com/lifeandstyle/2010/jun/02/brief-history-transgender-issues

Wirth-Cauchon, J. (2001). *Women and borderline personality disorder: Symptoms and stories*. New Brunswick, NJ: Rutgers University Press.

Yeginsu, C. (2018, February 15). Transgender woman breast-feeds baby after hospital induces lactation. *New York Times*. Retrieved from https://www.nytimes.com/2018/02/15/health/transgender-woman-breast-feed.html

Chapter 5

Brooks, R. (2017, April 5). "He," "She," "They" and Us. *New York Times*. Retrieved from https://www.nytimes.com/2017/04/05/insider/reporting-limits-of-language-transgender-genderneutral-pronouns.html

Butler, J. (1990). *Gender trouble.* New York, NY: Routledge.

Cutler, S. (2015, January 16). Sexist Job titles and the influence of language on gender stereotypes. *Brigham Young University Humanities*. Retrieved from http://humanities.byu.edu/sexist-job-titles-and-the-influence-of-language-on-genderstereotypes/

Erickson-Schroth, L. (2015). Psychological and biological influences on gender roles. In D. W. Pfaff & N. D. Volkow (Eds.), *Neuroscience in the 21st century* (pp. 1–22). New York, NY: Springer.

Garcia, S. (2018, November 27). Independent autopsy of transgender asylum seeker who died in ICE custody shows signs of abuse.

New York Times. Retrieved from https://www.nytimes.com/2018/11/27/us/trans-woman-roxsana-hernandez-ice-autopsy.html

Gay, V., Hicks, D., & Santacreu-Vasut, E. (2016, September 10). What languages can teach us about gender norms of behaviour. *VOX CEPR Policy Portal*. Retrieved from https://voxeu.org/article/languages-and-gender-norms-behaviour

Gumperz, J. J., & Levinson, S. C. (1996). *Rethinking linguistic relativity*. Cambridge, England: Cambridge University Press.

Hackett, C., & McClendon, D. (2017, April 5). Christians remain world's largest religious group, but they are declining in Europe. *Pew Research Center*. Retrieved from http://www.pewresearch.org/fact-tank/2017/04/05/christians-remain-worlds-largest-religious-group-but-they-are-declining-in-europe/

Hersher, R. (2013, April 24). "Yo" Said What? *National Public Radio*. Retrieved from https://www.npr.org/sections/codeswitch/2013/04/25/178788893/yo-said-what

Jacobi, T., & Schweers, D. (2017, April 11). Female Supreme Court justices are interrupted more by male justices and advocates. *Harvard Business Review*. Retrieved from https://hbr.org/2017/04/female-supreme-court-justices-are-interrupted-more-by-male-justices-and-advocates

Krulwich, R. (2009, April 6). Shakespeare had roses all wrong. *NPR*. Retrieved from https://www.npr.org/sections/krulwich/2009/04/06/102518565/shakespeare-had-roses-all-wrong

Lakoff, R. (1973). Language and woman's place. *Language in society*, 2(1), 45–79.

McDonald, C. (2015). *Captive genders: Trans embodiment and the prison industrial complex*. Edinburgh, Scotland: AK Press.

National Center for Victims of Crime. www.rainn.org/statistics/children-and-teens, https://www.d2l.org/wp-content/uploads/2017/01/all_statistics_20150619.pdf, https://www.parentsformeganslaw.org/statistics-child-sexual-abuse/

National Coalition Against Domestic Violence. (n.d.). Statistics. Retrieved from https://ncadv.org/statistics

Neuman, S. (2015, March 27). He, she or hen? Sweden's new gender-neutral pronoun. *National Public Radio*. Retrieved from https://www.npr.org/sections/thetwo-way/2015/03/27/395785965/he-she-or-hen-sweden-s-new-gender-neutral-pronoun

Owen, T. (2018, February 1). That ICE prison for transgender immigrants? It never opened. *Vice News*. Retrieved from https://news.vice.com/en_us/article/7xegnz/that-ice-prison-for-transgender-immigrants-it-never-opened

Panko, B. (2016, December 2). Does the linguistic theory at the center of the film "Arrival" have any merit? *Smithsonian Magazine*. Retrieved from http://www.smithsonianmag.com/science-nature/does-century-old-linguistic-hypothesis-center-film-arrival-have-any-merit-180961284/

Petrosky, E., Blair, J. M., Betz, C. J., Fowler, K. A., Jack, S. P., & Lyons, B. H. (2017). Racial and ethnic differences in homicides of adult women and the role of intimate partner violence—United States, 2003–2014. *Morbidity and Mortality Weekly Report*, *66*(28), 741.

Pew Research Center. (n.d.). Religious landscape study. Retrieved from http://www.pewforum.org/religious-landscape-study/#religions

Pew Research Center. (2015, May 12). America's changing religious landscape. Retrieved from http://www.pewforum.org/2015/05/12/americas-changing-religious-landscape/

Sanghani, R. (2017). Feisty, frigid and frumpy: 25 Words we only use to describe women. *The Telegraph*. Retrieved from https://www.telegraph.co.uk/women/life/ambitious-frigid-and-frumpy-25-words-we-only-use-to-describe-wom/

Schroth, P. W., Erickson-Schroth, L., Foster, L. L., Burgess, A., & Erickson, N. S. (2018). Perspectives on law and medicine relating to transgender people in the United States. *American Journal of Comparative Law*, *66*(Suppl 1), 91–126.

Sirimanne, C. R. (2016). Buddhism and women: The Dhamma has no gender. *Journal of International Women's Studies*, *18*(1), 273–292.

Zimmer, B. (2009, October 23). On language: Ms. *New York Times*. Retrieved from https://www.nytimes.com/2009/10/25/magazine/25FOB-onlanguage-t.html

Chapter 6

Anzaldúa, G., & Keating, A. (Eds.). (2013). *This bridge we call home: Radical visions for transformation*. New York, NY: Routledge.

binaohan, B. (2014). *decolonizing trans/gender 101*. Toronto, ON: Biyuti.

Bornstein, K. (1997). *My gender workbook: How to become a real man, a real woman, the real you, or something else entirely*. New York, NY: Routledge.

Bornstein, K., & Bergman, S. B. (2010). *Gender outlaws: The next generation*. Berkeley, CA: Seal Press.

Boston Women's Health Book Collective. (1973). *Our Bodies, Ourselves: A Book By and For Women*. New York, NY: Simon and Schuster.

Brittan, A. (1989). *Masculinity and power*. New York, NY: Blackwell.

Butler, J. (2002). *Gender trouble*. New York, NY: Routledge.

Califia, P. (1997). *Sex changes: The politics of transgenderism*. San Francisco. CA: Cleis.

Chu, A. L. (2018, November 24). My new vagina won't make me happy. *The New York Times*. Retrieved from https://www.nytimes.com/2018/11/24/opinion/sunday/vaginoplasty-transgender-medicine.html

Chu, A. L. (2019). *Females*. London: Verso.

Counter, R. (2012, August 17). Caitlin Moran forces us to ask, is it time for fifth-wave feminism? *The Globe and Mail*. Retrieved from https://www.theglobeandmail.com/arts/books-and-media/book-reviews/caitlin-moran-forces-us-to-ask-is-it-time-for-fifth-wave-feminism/article4486313/

Crenshaw, K. (1989). Demarginalizing the intersection of race and sex: A black feminist critique of antidiscrimination doctrine, feminist theory and antiracist politics. *University of Chicago Legal Forum, 1989*, 8.

Currah, P., Juang, R. M., & Minter, S. (Eds.). (2006). *Transgender rights*. Minneapolis: University of Minnesota Press.

Douglass, R. (2017, August 4). More men should learn the difference between masculinity and toxic masculinity. *HuffPost*. Retrieved from https://www.huffpost.com/entry/the-difference-between-masculinity-and-toxic-masculinity_b_59842e3ce4b0f2c7d93f54ce

Ellis, E. (2019, July 10). Reddit's 'manosphere' and the challenge of quantifying hate. *WIRED*. Retrieved from https://www.wired.com/story/misogyny-reddit-research/

Enke, A. (2012). *Transfeminist Perspectives in and Beyond Transgender and Gender Studies*. Philadelphia: Temple University Press.

Erickson-Schroth, L. (Ed.). (2014). *Trans bodies, trans selves: A resource for the transgender community*. New York, NY: Oxford University Press.

Faderman, L. (1981). *Surpassing the love of men: Romantic Friendship and Love Between Women from the Renaissance to the Present*. London: Women's Press.

Feinberg, L. (1996). *Transgender warriors: Making history from Joan of Arc to Dennis Rodman*. Boston, MA: Beacon Press.

Flexner, E., & Fitzpatrick, E. F. (1996). *Century of struggle: The woman's rights movement in the United States*. Cambridge, MA: Harvard University Press.

Foucault, M. (1990). The history of sexuality: An introduction (R. Hurley, trans.; Vol. 1). New York, NY: Vintage.

Gossett, R., Stanley, E. A., & Burton, J. (2017). *Trap door: Trans cultural production and the politics of visibility*. Cambridge, MA: MIT Press.

Grady, C. (2018, March 20). The waves of feminism, and why people keep fighting over them, explained. *Vox*. Retrieved from https://www.vox.com/2018/3/20/16955588/feminism-waves-explained-first-second-third-fourth

Halberstam, J. (2019). *Female masculinity*. Durham, NC: Duke University Press.

Hathaway, J. (2014, October 10). What is Gamergate, and why? An explainer for non-geeks. *Gawker*. Retrieved from http://gawker.com/what-is-gamergate-and-why-an-explainer-for-non-geeks-1642909080

Holland, S. P., & Cohen, C. J. (2005). *Black queer studies: A critical anthology*. Durham, NC: Duke University Press.

Jagose, A. (1996). *Queer theory: An introduction*. New York, NY: NYU Press.

Koyama, E. (2003). The transfeminist manifesto. In R. Dicker & A. Piepmeier (Eds.), *Catching a wave: Reclaiming feminism for the 21st century* (pp. 244–259). Boston, MA: Northeast University Press.

Koyama, E. (2006). Whose Feminism is it anyway? The Unspoken Racism of the Trans Inclusion Debate. In S. Stryker & S. Whittle (Eds.), *The transgender studies reader* (pp. 698–705). New York, NY: Taylor & Francis.

Kupers, T. A. (2005). Toxic masculinity as a barrier to mental health treatment in prison. *Journal of Clinical Psychology, 61*(6), 713–724.

Moraga, C., & Anzaldúa, G. (Eds.). (1981). *This bridge called my back: Writings by radical women of color*. New York, NY: SUNY Press.

Napikoski, L. (2020, March 10). What is "the second feminist wave?" *ThoughtCo*. Retrieved from https://www.thoughtco.com/the-second-feminist-wave-3528923

Raymond, J. (1994). *The transsexual empire: The making of the she-male* (Athene Series). New York, NY: Teachers College Press.

Rios, C. (2015, June 24). Rebel girls: The 5 Moments in feminist history that jump-started the second wave. *Autostraddle*. Retrieved from https://www.autostraddle.com/

rebel-girls-the-5-moments-in-feminist-history-that-jump-started-the-second-wave-295024/

Rizki, C., Rodríguez, J. M., Galarte, F. J., Correa, M. B., Estalles, C., Pericles, C., . . . Valencia, S. (Eds.). (2019). *Special issue: Trans Studies en las Américas. TSQ: Transgender Studies Quarterly, 6*(2).

Rubin, G. (1992). Thinking Sex: Notes for a Radical Theory of the Politics of Sexuality. In C. S. Vance (Ed.), *Pleasure and Danger: Exploring Female Sexuality* (pp. 267–293). London: Pandora.

Serano, J. (2007). *Whipping girl: A transsexual woman on sexism and the scapegoating of femininity.* Berkeley, CA: Seal Press.

Scott-Dixon, K. (2006). *Trans/Forming Feminisms: Transfeminist Voices Speak Out.* Toronto: Sumach Press.

Sheber, V. (2017, December 16). Feminism 101: What are the waves of feminism?. *FEM.* Retrieved from https://femmagazine.com/feminism-101-what-are-the-waves-of-feminism/

Snorton, C. R. (2017). *Black on both sides: A racial history of trans identity.* Minneapolis, MN: University of Minnesota Press.

Stanley, E., & Smith, N. (2015) *Captive genders: Trans embodiment and the prison industrial complex.* Edinburgh, Scotland: AK Press

Stone, S. (1992). The *Empire* strikes back: A posttranssexual manifesto. *Camera Obscura, 10*(2), 150–176.

Stryker, S. (2008). *Transgender history.* Berkeley, CA: Seal Press.

Stryker, S., & Aizura, Aren. (2013). *The transgender studies reader* (Vol. 2). New York, NY: Taylor & Francis.

Stryker, S., & Bettcher, T. M. (2016). Introduction: Trans/feminisms. *TSQ: Transgender Studies Quarterly, 3*(1–2), 5–14.

Stryker, S., & Whittle, S. (Eds.). (2006). *The transgender studies reader* (Vol. 1). New York, NY: Taylor & Francis.

Sycamore, M. (2006) *Nobody passes: Rejecting the rules of gender and conformity.* Berkeley, CA: Seal Press.

Thom, K. (2018, November 29). The pain—and joy—of transition. *Slate.* Retrieved from https://slate.com/human-interest/2018/11/andrea-long-chu-new-york-times-criticism-response-transgender.html

Tolentino, J. (2018, May 15). The rage of the incels. *The New Yorker.* Retrieved from https://www.newyorker.com/culture/cultural-comment/the-rage-of-the-incels

Wilchins, R. A., & Serano, J. (1997). *Read my lips: Sexual subversion and the end of gender.* Ithaca, NY: Firebrand Books.

Winton, T. (2018, April 9). About the boys: Tim Winton on how toxic masculinity is shackling men to misogyny. *The Guardian*. Retrieved from https://www.theguardian.com/books/2018/apr/09/about-the-boys-tim-winton-on-how-toxic-masculinity-is-shackling-men-to-misogyny

Xavier, J. (1999). Passing as privilege. Retrieved from https://learningtrans.files.wordpress.com/2010/11/jxavier_passing_as_privilege.pdf

Chapter 7

Atwood, M. (1985). *The Handmaid's Tale.* Toronto: McClelland and Stewart.

Barry, E. (2018, March 24). In Sweden's preschools, boys learn to dance and girls learn to yell. *New York Times*. Retrieved from https://www.nytimes.com/2018/03/24/world/europe/sweden-gender-neutral-preschools.html

Calkin, S. (2018). Abortion access is changing through technology and activism. *Discover Society*. Retrieved from https://discoversociety.org/?s=calkin.

Cima, R. (2017). Bias, she wrote. The gender balance of The New York Times best seller list. *The Pudding*. Retrieved from https://pudding.cool/2017/06/best-sellers/

Economic inequality across gender diversity. (n.d). *Inequality.org*. Retrieved from https://inequality.org/gender-inequality/#gender-income-gaps

Geiger, A., & Kent, L. (2017, March 8). Number of women leaders around the world has grown, but they're still a small group. *Pew Research Center*. Retrieved from https://www.pewresearch.org/fact-tank/2017/03/08/women-leaders-around-the-world/

Gould, L. (1972). *X: A fabulous child's story.* http://chawkinson.pbworks.com/w/file/fetch/88063807/A%20Fabulous%20Child%27s%20Story.pdf

Madrigal, A. (2014, June). Making babies. *The Atlantic*. Retrieved from https://www.theatlantic.com/magazine/archive/2014/06/making-babies/361630/

Morris, A. (2018, April 2). It's a theyby! Is it possible to raise your child entirely without gender from birth? Some parents are trying. *New York Magazine*. Retrieved from https://www.thecut.com/2018/04/theybies-gender-creative-parenting.html

Oxfam International. (n.d). Why the majority of the world's poor are women. Retrieved from https://www.oxfam.org/en/even-it/why-majority-worlds-poor-are-women

Scientists a step closer to mimicking way human body creates sperm. (2018, January 1). *The Guardian*. Retrieved from https://www.theguardian.com/science/2018/jan/01/scientists-a-step-closer-to-mimicking-way-human-body-creates-sperm

UN Women. (2012). Facts and figures. Retrieved from http://www.unwomen.org/en/news/in-focus/commission-on-the-status-of-women-2012/facts-and-figures

UNESCO. (2013). Education for All Global Monitoring Report: Fact sheet. Retrieved from https://en.unesco.org/gem-report/sites/gem-report/files/girls-factsheet-en.pdf

Women in academia: Quick take. (n.d.). *Catalyst*. Retrieved from https://www.catalyst.org/knowledge/women-academia

Zhang, S. (2017, December 21). A woman gave birth from an embryo frozen for 24 Years. *The Atlantic*. Retrieved from https://www.theatlantic.com/science/archive/2017/12/frozen-embryo-ivf-24-years/548876/

INDEX

For the benefit of digital users, indexed terms that span two pages (e.g., 52–53) may, on occasion, appear on only one of those pages.